Building an Effective IoT Ecosystem for Your Business

Sudhi R. Sinha • Youngchoon Park

Building an Effective IoT Ecosystem for Your Business

Foreword by Scott Guthrie

Sudhi R. Sinha
Johnson Controls International
Milwaukee, Wisconsin, USA

Youngchoon Park
Johnson Controls International
Milwaukee, Wisconsin, USA

ISBN 978-3-319-86151-7 ISBN 978-3-319-57391-5 (eBook)
DOI 10.1007/978-3-319-57391-5

Printed on acid-free paper

This Springer imprint is published by Springer Nature
The registered company is Springer International Publishing AG
The registered company address is: Gewerbestrasse 11, 6330 Cham, Switzerland

To my lovely better half Ms. Sohini Sengupta—the pillar behind all my adventures and successes. Without your support, encouragement, and patience, none of this would have taken shape.

– Sudhi

To my dear wife Ms. Jungeun Kim and my children Hannah Park and Eugene Park—you are the force behind my success.

– Dr. Youngchoon Park

Foreword

https://news.microsoft.com/exec/scott-guthrie/#bjRibuwYG4LSbv41.97

The Internet of Things (IoT) represents an opportunity to drive a step-change improvement in what continues to be the core mission of every company: deliver great products and delight customers. IoT enables continuous learning from your customers (and how they use your products) that eventually will give companies sustainable advantage. At the heart of this process lies the creation of a Digital Feedback Loop, which is only possible if signals from customers, employees, and operations are digital. As such, every enterprise must embark on Digital Transformation to succeed in this new paradigm. IoT is emerging as the foundation for the digital feedback loop due to the confluence of the continuation of Moore's law to provide affordable and progressively advancing compute technology and step changes in communication technology.

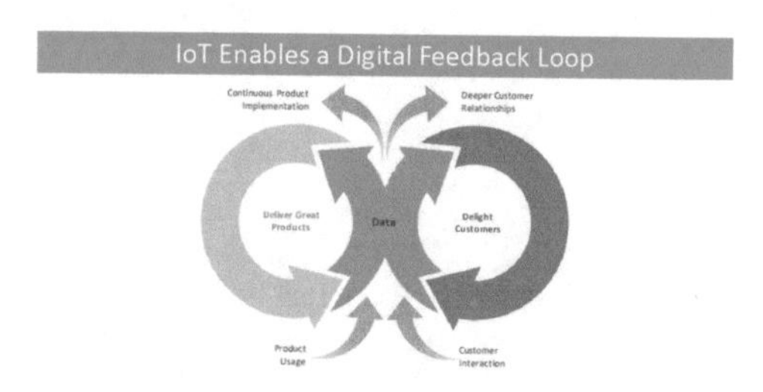

Sudhi Sinha and Dr. Youngchoon Park are close partners to Microsoft in their official capacity at Johnson Controls and have been working on enabling the digital feedback loop as a key element of the digital transformation for Johnson Controls with advanced projects such as connected chillers and others. Their book is written by practitioners that have embarked on this critical journey for an established and highly valued company. The lessons learned and their experiences will help you on your journey to drive digital transformation to deliver great products and delight your customers.

Microsoft Cloud and Enterprise Group — Scott Guthrie

Acknowledgments

We feel incredibly lucky to have worked on this project, our labor of love. This is a milestone in our quest for learning about and furthering the space of IoT, which we think will fundamentally transform the way we live and interact in this world. There are many people who have helped us in this journey, and without their contribution, none of this would have come to life.

We would first like to thank Mr. Bill Jackson, President of Global Products in Johnson Controls, who encouraged us, challenged us, and supported us for this project. You make everybody who comes in your contact a better person and professional. We would also like to thank Uli Homman, Distinguished Architect at Microsoft, for being a major supporter for us. We thank Karl Reichenberger, Director of IP and Legal for Johnson Controls, who painfully went through the material to ensure that we have the right IP protection and confidentiality for Johnson Controls; Karl was also a key input provider to our work; he is not only a brilliant lawyer but an excellent engineer as well. We thank our colleagues in the academia Prof. AnHai Doan of UW-Madison, Prof. Subbarao Kambhampati and Prof. Selcuk Candan of Arizona State University, Prof. Jignesh Patel of UW-Madison, Prof. Mario Berges of Carnegie Mellon University, and Prof. Alexis Abramson of CWRU for the discussions that helped shape many of our thoughts. We express our gratitude to Ms. Susan Lagerstrom-Fife of Springer and her associate Ms. Sasikumar Sharmila from SPi Global for giving us valuable inputs and keeping us on track. Last but not least, we thank our publisher.

Contents

Chapter 1
Building an IoT Ecosystem Framework

"You don't have to be a genius or a visionary or even a college graduate to be successful. You just need a framework and a dream."

—Michael Dell

Transformative technologies have historically revolutionized the way industry works, not because a single new technology is created, but rather because consensus around a group of existing and emerging technologies emerges that provides a cohesive way to think about business innovation. We are now at the cusp of such a transformation with technologies around the Internet of Things (IoT) [1–3].

Surprisingly, the raw ingredients that go into composing the DNA of IoT include the very things that many industries have practiced at smaller scales and higher costs for decades. These ingredients are embedded sensors, networking, connected components, cloud computing, data analytics, and reactive services. Arguably, many industry leaders have silently pioneered many of the core technologies and innovative business models that are needed to take controls into the IoT era. There are significant additional opportunities for both technological and business growth by building on connected products and services. One way to cast this future vision is to think of evolving business models, establishing new customer relationships, galvanizing organizational change, and creating new technological core competencies in reshaping the business based on connected products, services and collected data along the life-cycle of products [4–6].

In this chapter, we present the IoT framework that represents the key components and technologies to deliver an end-to-end solution for connected products and services. In order to understand the framework, we will first look at the generalized view of IoT from the platform perspective, then the applied framework specific to industrial equipment service management. This chapter will form the basis of the next few chapters where each of the components will be explained in great details.

S.R. Sinha, Y. Park, *Building an Effective IoT Ecosystem for Your Business*,
DOI 10.1007/978-3-319-57391-5_1

1.1 IoT Ecosystem Development Framework

As IoT evolves and enables completely new markets, we need to create an ecosystem of alliance partners to help us get to market faster, share risks and create win-win situations with them. We need to understand where we can bring value and where we need help.

Authors believe that IoT will change a way to do business today, current business value chain in a market and perceived values from connected products and services will change. Traditional transaction-based product business will be transformed into life-cycle management. An IoT ecosystem must be able to capture and model them as they evolve during the product lifecycle [6, 7].

We should be able to capture every event that a connected product will generate during its life cycle. Each event may lead different values for OEM, a service provider, a user and other involved entities. IoT ecosystem must able to model such value distribution dynamics and must provide value protection mechanism (e.g., access control and authorization) for parties who are in the value distribution.

As part of the ecosystem design, we have developed a referenceable framework that helps technology partnering strategy map (see Fig. 1.1). Each vertical represents technology components that could be developed in-house, purchased from or co-developed with a vendor or partner.

1.1.1 Smart Connected Things

A smart connected thing or smart device is defined as one which has communication capabilities, compute power, and can make decisions at a local level in a limited context [8]. Communication can be through any wireless or wired mechanism. Typically, wireless methods are preferred because it eliminates the cost of wiring

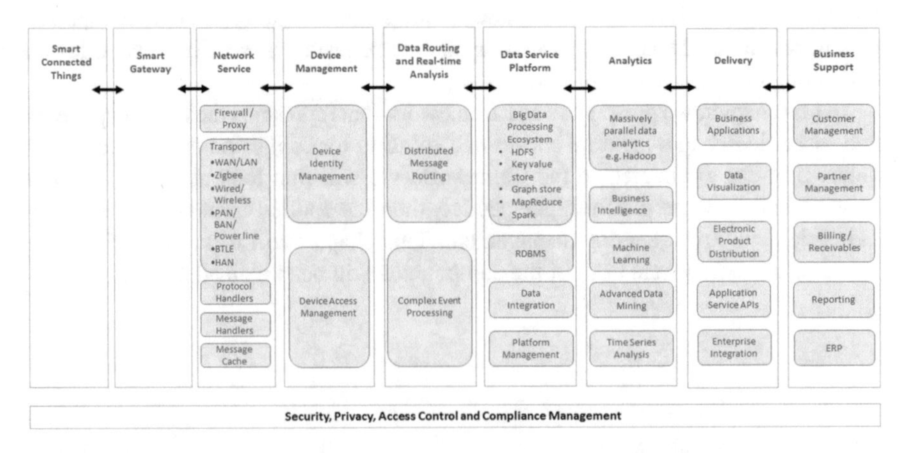

Fig. 1.1 A proposed IoT ecosystem development framework

and labor associated with it; unless the devices are leveraging already established wired communication methods like any present IP backbone. Compute power is enabled by a microprocessor chip with an embedded operating system like Linux, Android, iOS, RTOS, Windows 10 IoT. The computing power is used to store data locally and temporarily in the device and execute pre-determined decision trees. Increasingly we are seeing local level limited isolated fault detection and diagnostics being performed within the devices. Since there is an abundance of computing power in the devices, basic fault response capabilities might make the device more adaptive and responsive to changing environmental/operational conditions [9].

They are things like LED lights, mobile phones, smart street lamps, cars, televisions, and actuators with embedded intelligence and communication capabilities. They connect to other smart things, and remote services. Connected products allow the OEM to know more about how customers use its products, to broaden the value proposition beyond the physical product, to include valuable data and value-added services, and form closer relationships with customers. Manufacturers must consider multiple connection types while developing connected products. A product could be connected to a home, an infrastructure, an owner, OEM, another service provider, an enterprise and other smart connected things [6, 10]. Figure 1.2 illustrates a software, more specifically an operating system level, an architecture for a smart thing by [11].

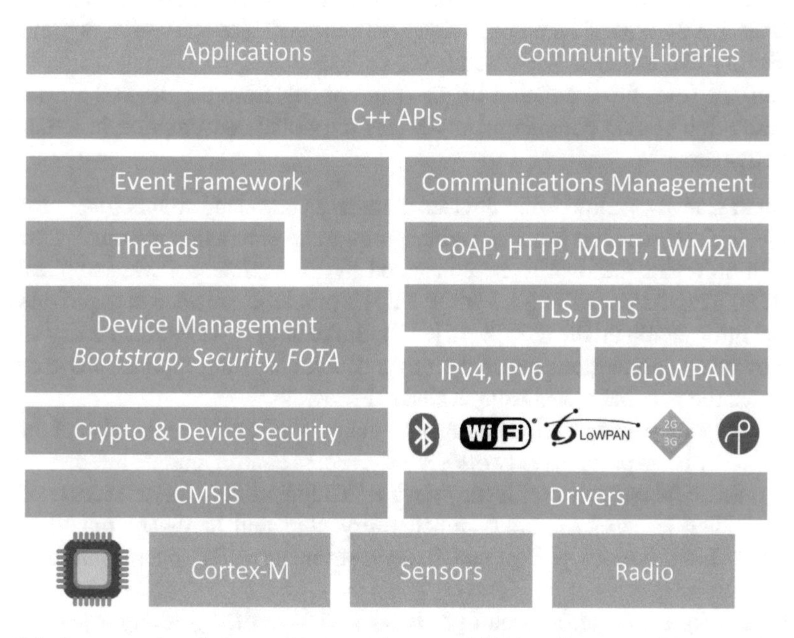

Fig. 1.2 An example of Smart thing architecture. ARM's reference architecture for IoT development

When devices are not smart but need to be part of an IoT ecosystem, we see device or field gateways being used as a proxy to allow communication over public or private networks using fully secure defensible server platforms with high capability hardware. We will further discuss gateways in the next section.

1.1.2 Aggregate and Enrich: Gateway

The IoT will include not just modern IoT-enabled devices but also systems that are already in place today and operate outside of the cloud-based IoT solutions such as legacy controls systems. Those legacy devices that are not IP-based e.g., still most field control devices in manufacturing industries are based on OPC [12], Modbus [13], and BACnet [14] that are based on non-Internet based to connect cloud services or other devices without full Ethernet or Wi-Fi interface. In order to build applications on top of those protocols, these protocols have to be converted into Internet Protocol (IP). The IoT gateway is a soft component or a physical device with embedded software components that performs such translation.

A typical gateway solution has many functions:

1. Discover and communicate with legacy devices through non-internet based protocols
2. Translate and transport acquired data and command via Internet protocol
3. Optimize the cost of communication between the device and rest of digital world
4. Identify the resources contained in the devices
5. Perform semantic mediation of the data coming from the devices i.e. naming translation of data obtained from devices for global consumption in a standardized manner

Smart gateways allow smart devices to make an outbound connection to a public or private cloud which is used for further storage and processing in your IoT ecosystem. In order for the data to be consumed by the various software applications, RESTful API, CoAP, HTTP(s), AMQP, MQTT protocols are needed to provide contextual information on the data [8, 15]. APIs are definitely widespread today and are used by many corporations to use their own data for building applications and share data with the external world using a clean abstraction.

Not only supporting various transmission and application protocols but also, an IoT gateway often provides network security, access control, unique internet address (e.g., a RESTful endpoint) of legacy devices' endpoints for remote access and protocol mediation services. Pretty much every chip and hardware manufacturers including Intel, Freescale, Dell and Texas Instruments offer general-purpose gateway solutions.

In their implementation, smart gateways will have field gateway aspects which primarily deal with the devices and cloud gateway aspects that enable high-scale telemetry ingestion, device identification, device management, and cloud-to-device commanding (when allowed). When there is a high density of devices in a location, we have seen one or many field gateways connecting through one cloud gateway. In your home, several smart devices like the security system, TV, refrigerator, music systems, printers, computers, washing machine, dishwasher, coffee maker, etc. could have elements of field gateways built into them and you could use a singular home automation device to act as the cloud gateway. In a manufacturing or process plant, individual production line machines will typically have field gateways built into them and connect via some common cloud gateway organized by production lines, processes or zones. Similarly in an office building, for the HVAC system, there will be dozens of devices and hundreds of sensors with smart field gateway capabilities connected through a common cloud gateway. These are just some of the examples of how IoT devices will connect. Several device manufacturers are building direct cloud connectivity and will try to position their own cloud solution for data and analytics. You need to make decisions based on at what level of aggregation does your business or needs get best served and who can solve your needs best—you or a particular solution provider. If you are sourcing IoT devices from multiple places, look for easy connectivity to your platform; if you are a device manufacturer, make connectivity to your device easy to make it attractive for wide adoption.

In summary, a gateway can be considered as a network router (i.e., bridging the intranet and the internet) with a legacy and IoT-enabled device integration device to enable internet-based data exchange as well as to share wide area network connectivity. It must support flexible deployment topologies to accommodate various customers' network constraints and support open IoT protocols such as CoAP, XMPP, AMQP, MQTT, and web based common data exchange such as HTTP RESTful APIs to meet standard requirements (e.g., smart home and light commercial markets are heavily driven by these standards) (Fig. 1.3).

1.1.3 Network Services

These are services that execute the transport of data from devices to the public or private cloud. Network services also help with the translation of different messaging protocols between different devices and communication networks. The network services include enabling and managing the fixed or wireless communication with the smart things. Here the vendors include traditional telecommunication service providers such as Verizon, T-mobile, and AT&T, as well as newcomers such as Google Fiber, are providing data transmission management services. Often, the connected device needs multiple device provisioning and registration steps to establish a connectivity via wide area networks. First, a smart thing needs to be provisioned and registered into a telecommunication service provider's network if it has embedded-SIM or Embedded Universal Integrated Circuit Card to establish Internet connectivity. Upon a successful provisioning to telco's system, a smart thing can register itself to a target application (e.g., manufacture's cloud service).

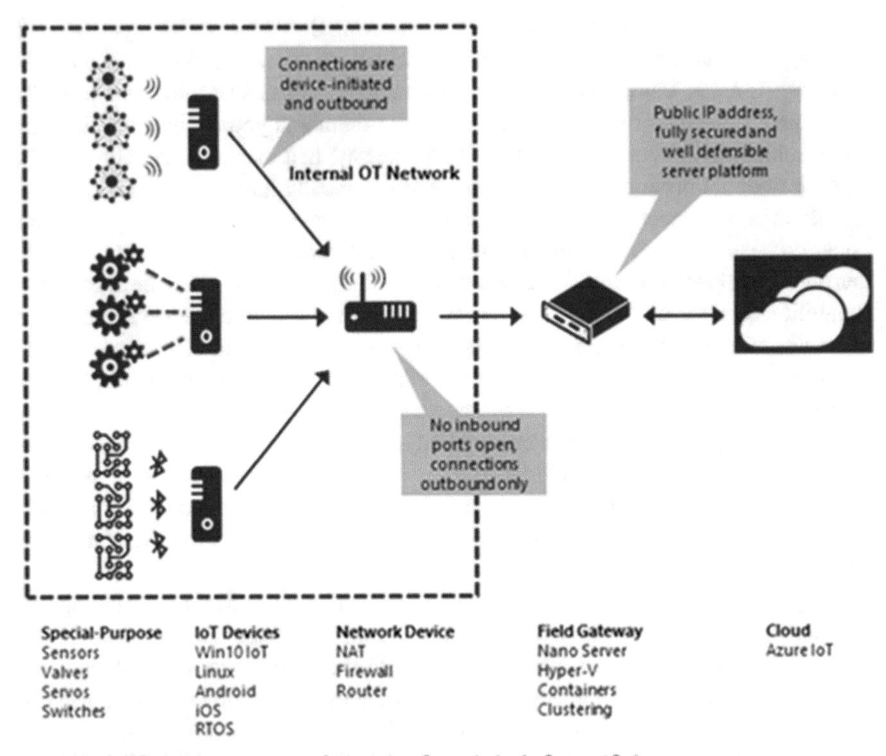

Fig. 1.3 IoT connectivity architecture

There are multiple communication protocols which might get used based on your products and context. Communication could be established over direct wired connections or wireless methods between devices and network. The popular wireless communication protocols Wi-Fi, Bluetooth, Zigbee, and EnOcean [15]. There are different types of networks which are established—Near Field Communication (NFC), Body Area Network (BAN), Personal Area Network (PAN), Home Area Network (HAN), Wide Area Network (WAN), Local Area Network (LAN), Internet Area Network (IAN) for direct cloud connectivity [16, 17].

The protocol handlers will facilitate the translation received over these multiple protocols into common understandable and usable formats; message handlers take care of the transport of the messages, caching enables faster retrieval of data. The firewall and proxy components will drive your security and privacy policies. The following diagram depicts the device to cloud connectivity by leveraging devices, gateways and network services:

When you are run your operation globally, contract and billing management of telecommunication services becomes an operational challenge, where we should have a global operational capability as one of the partnership evaluation criteria.

Software integration with telecommunication service provider's real-time device and data usage must be considered in a connected product development.

To advocate and defend "every product and service provider must have network security story", it is highly desired to build a partnership with few major network carriers. Many network carriers pro-vide a secure connection option as a part of premium services, however, a similar degree of network securities can be achieved via employing trust platform chip in your connected products and standard encrypted messaging such as AMQP via IoT secure transport.

Multiple protocol support should be a part of your strategy that enables flexibility in connecting more devices and services. For example, starting from the sensor transmitting data over the air, the popular transport layer protocols are WiFi, Bluetooth, La3G/4G, 802.15.4 and Zigbee [17–19]. Each of them has different characteristics and they probably will stay in the market for a long time. The key tradeoff is power vs. bandwidth and what types of use cases the company wants to build.

1.1.4 Device Management

The biggest transformative aspect about IoT is the multitude and variety of devices now getting connected to create new opportunities and insights. Managing these devices requires special focus, infrastructure, and robust policies. There are four key aspects of managing these myriad devices both from an identity and access management:

1. **Device provisioning**—Setup of the device to establish a secure trusted relationship with the digital world. By generating a secure key and assigning it to a device, the device will be ready to interact with the rest of IoT ecosystem.
2. **Device registration**—Maintaining a list of provisioned devices and associated metadata.
3. **Telemetry ingestion**—Providing a simple, secure path for receiving data from devices.
4. **Command and control**—Providing functionality for securely sending data to a device, while using existing outbound connections.

For complete lifecycle management of the devices, there are several actions involved like:

1. Tracking devices from manufacturing to install
2. Distributing and licensing device contained software including remote updates for new features, security, other maintenance type activities
3. Recovering and decommissioning of non-usable devices post useful life
4. Monitoring device health
5. Learning about customer device usage patterns, release levels, and maintenance protocols to extract additional business value

We see different types of business models around devices, connectivity and data analytics from devices—device as a service, platform as a service, technology framework led solutions and device sales amongst others.

In terms of device management implementation, we see two major approaches—one where the edge devices are smart meaning there are adequate electronics for processing power and outbound connectivity built into the device itself, and the other where not so smart devices need gateway devices for greater processing capabilities and outbound telemetry connection. Interface agents in these scenarios have several requirements:

1. Secure end-to-end connectivity
2. Easy integration into embedded products
3. Simple registration workflow
4. Easy diagnosis and remediation of connectivity problems
5. Remote user access for operation (with ability to override at LAN)
6. Fast and reliable software update
7. Time series data push
8. Remote write-backs to device via automation
9. Device location tracking
10. Device time sync (consider time zone and automated changes to time zone in events like daylight savings)
11. Licensing of device and features within the device
12. Rules engine/scheduling of device actions from cloud
13. Event notification
14. API's for extension of capabilities in devices, and in the cloud

Device management is one of the most involved and rapidly evolving aspects of IoT. Almost every organization has some in-house capabilities in this space. Evaluate your options on how best to leverage internal capabilities and extend them organically or engage in outsourced partnerships to build stronger and expansive IoT framework components around device management.

1.1.5 Data Routing and Real-Time Analysis

There are two major aspects of data routing and real-time analysis—distributed message routing and complex event processing. Distributed message routing refers to sending the messages and data elements generated from the devices to the right data platform storage locations. Generally, in IoT environments, real-time message hub (e.g., Kafka, Storm or Event Hub) act as the front door for event ingestion and sometimes for also bi-directional control action conduit to the devices or system from which the message events are received [20, 21]. The storage locations or applications use listener applications to receive and process the messages. Event Hubs act as the backend distribution infrastructure [22].

Complex event processing (CEP) also known as hot path data analytics refers to identifying meaningful events and responding to them quickly. Anomalies or other interesting patterns trigger an event in the system that can start additional processing or logic.

There are multiple steps involved—data generation from event producers, a collection of the data, queuing of events, the transformation of events, preparation for and actual long term storage of the events, and finally the multi-format presentation of the analyzed events for reporting or further action. Think of complex event processing as a mini self-contained special purpose IoT ecosystem. There are many open source and licensed solutions available for complex event processing. Following diagram depicts the Apache Storm open source solution for CEP (Fig. 1.4):

In traditional data management and enterprise systems, message routing and event processing types of activities are sequential and spaced by more elaborate classification, cleansing, storage and analytical steps. In the IoT world, they are almost simultaneous events because we want to take the advantage of early insights from time series data to take immediate remedial actions. This is the first stage in the framework to take advantage of elementary machine learning. Today we have solutions like Apache Storm which can process millions of events per second real-time in different computing environments to meet the needs of real-time analysis. Distributed message routing also involves ingestion, processing and placing back millions of messages in message distributor for multiple listener applications to pick up. Sometimes during this activity, we see the occurrence of tasks like data transformation and validation.

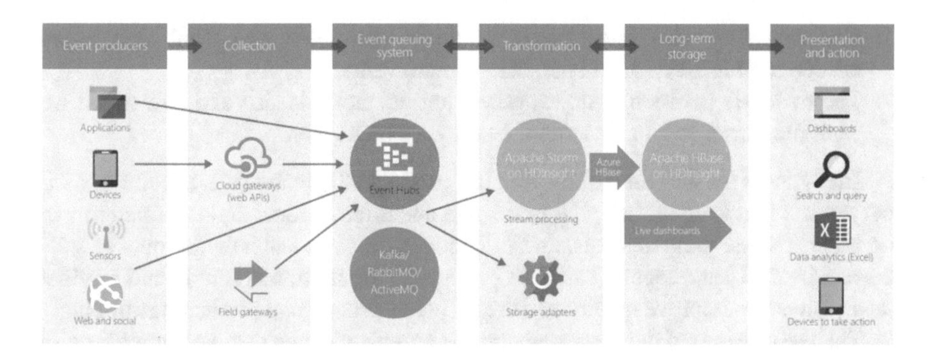

Fig. 1.4 Apache Storm open source solution

1.1.6 IoT Data Platform

The IoT platform is the networking, messaging, data management and processing infrastructure which includes analytics and modeling, the cloud and IT network, all in one packaged component. This abstraction is to ensure interoperability within products and among other products. All of them must operate on the same platform. There are quite a number of vendors on the market that are offering IoT cloud/platform solutions and picking one that fits the company vision and use cases will help accelerate its efforts in IoT.

Many large IT solution providers such as Microsoft, Intel, Cisco, Google, Amazon, SAP, Qualcomm, and Oracle offer general-purpose solutions with developer tools for customization, while ThingWorx, Axeda (now a part of PTC), ETHERIOS, and GE offers more vertically integrated industry solutions such as manufacturing sectors [6].

The key question in making platform selection is—"what is the data needed to transform the business and what do we do with the model?" An important aspect of an IoT product is that a significant portion of the value will be generated from the software version of the physical product. Models are created to abstract the functionality of the physical product and analytics should exist for the part, product and system level. We can then build applications using the models to generate value. Even a simple app that changes a few key parameters and attributes of the product e.g. temperature and air flow, can be quite valuable. The software version of the product is the hearts and the minds of the IoT product that will derive IoT platform requirements.

Operational Technology has to marry with Information Technology to bring IoT-enabled products to life. Today, there are an increasing number of products that have embedded sensors, actuators and systems on them that provide valuable data about the product. The company has to determine the capabilities of the sensor being put on the product and the software support needed e.g. API, setup, installation and closed-loop updates. The decision is mainly driven by

1. Understanding what information we don't have today and what sensors are needed to pull data that can help us generate value.
2. The tradeoffs between costs vs. power and memory needed and how much we want the sensor to be able to do the analytics on its own.

Analytics is the tool for building the IoT products. The essence is to build statistical models that keep on bringing more data about the product and infer patterns to do comparisons, trend analysis, and predictions, or even teach the product to self-correct itself. Usage reports can also be created for sales, marketing, and product development to improve quality and create better pricing and product positioning.

The use of external data can augment the company's own data and provide extra information for better decision making. Geolocation of customer address (e.g., latitude and longitude of customer aggress) and census data (e.g., income profile) may help to develop customer segmentation.

The IoT product is not going to realize its full value if it is not integrated well with the business systems namely the CRM, SCM, ERP and PLM components to adapt to changes in the business processes.

In the IoT ecosystems, we frequently see the convergence of structured, unstructured and streaming data. The amount of data generated by the hundreds, thousands and sometimes millions of devices becomes quite massive. This environment requires very robust data platform, and flexible and scalable data services to leverage the power of data. This is where a lot of the Big Data technologies come into play.

1.1.7 Data Service Platform

1. **Big Data processing ecosystem**—The Big Data processing ecosystem is the core of the Data Service Platform. Apache Hadoop is a collection of many tools and data management eco-system [23] from open source releases that deal with Big Data needs.

   The Hadoop Distributed File System (HDFS) executes distributed parallel storage using name nodes to track the placement of physical data across various Hadoop instances and data nodes which actually physically store the data. Multiple simple consumer grade computing devices can be organized virtually to store very large amounts of data using Hadoop.

   Hadoop MapReduce system is very useful in processing large volumes and transactions of data—structured, unstructured and streaming. MapReduce uses parallel distributed algorithms for orchestration of the clustering required for storage and processing of data.

   Data can have different meanings, this is driven by various interpretations of the data and insights. This requires us to have an abstraction of data into multiple meanings through a collection of keys and value pairs of the data.

   This is achieved through associative arrays. A key-value store; a type of NoSQL allows for managing associative arrays through dictionary or hash. Dictionaries contain a collection of objects, or records, which in turn have many different fields within them, each containing data. These records are stored and retrieved using a key that uniquely identifies the record, and is used to quickly find the data within the database. Traditional relational databases have predefined data structure in the database as a series of tables containing fields with well-defined data types. On the other hand, key-value systems treat the data as a single opaque collection which may have different fields for every record. This offers considerable flexibility and more closely follows modern concepts like object-oriented programming. Because optional values are not represented by placeholders as in most relational databases, key-value stores often use far less memory to store the same database, which can lead to large performance gains in certain workloads [24].

Graph databases or graph stores use graph structures for semantic queries with nodes (entities), properties (information about entities) and edges (connection between nodes and properties) to store data [25, 26]. This is a very time and resource efficient method to store associative data sets.

Spark [27] is an open source cluster computing framework built using multi-stage in-memory primitives. Spark accelerates processing performance by an order of magnitude and very well suited for machine learning typical of IoT environments.

There is extensive guidance provided on building your Big Data platform in the author's book titled "Making Big Data Work For Your Business"; chapter 4, deals specifically with the topic of the technology landscape.

2. **RDBMS**—Usually consumer-end or front-end applications have very low latency data retrieval needs from the data store. In general Big Data enables low latency. But if the data structures used by the consuming applications are very standard, it may be a good idea to have processed data required by those consuming applications be stored in an RDBMS, as RDBMS has natural low latency if the data is modeled well. When doing so, usually the data is parsed, transformed, enriched before being stored in RDBMS storage.
3. **Data Integration**—IoT relies on being able to collect and connect data from any and every device. There are three parallel forces which influence this—rapid evolution of communication technologies including protocols and methods, dynamic evolution overall IoT technology landscape, and finally, changes in the design/function/capabilities of the devices themselves. Data communication needs to mirror the purpose and functioning of the device, and not constrained by the need to make it easy for external usage needs. The data could be structured, unstructured or streaming. Formats of the data collected will be different owing to the communication protocols which runs into dozens in any given IoT ecosystem. We need to design our storage and processing in a way that the native state of data does not get destroyed for it might be used and analyzed differently later. At the same time, we also need the easy semantic mediation of the data for the data platform to use it meaningfully. Device manufacturers and technology providers always talk about APIs as the solution to being able for the world to interact with their device or utility; however, there is al-ways some level of proprietary adoption required which leads to more than advertised effort for seamless integration of data and devices. There are few generally accepted standards and broad consensus on this topic. This is so because, on one hand, the whole industry is still in relative infancy with new opportunities and use cases get-ting discovered every day; on the other hand, different participants in the industry are trying to influence a position to their advantage. People are also finding new ways of connecting, collecting and analyzing better every day, so there is constant flux. These factors make data integration acquire new levels of profound complexities in the world of IoT.

In building the data integration components for your IoT framework, keep in mind the following:

1. Collect and store the data from devices and other applications in as native state as possible. Use layers of abstraction for any processing and further storage
2. Use as much of asynchronous communication as possible, this will also help with the previous goal while making integration simpler
3. Use micro services to build your integration fabric
4. In your technology landscape, do not get over reliant on any particular integration software or technique, this will limit your future flexibility
5. Make integrations database independent
6. Platform Management

In the IoT framework context, platform management includes a broad set of activities like device management, network and communication management, data management, hosting management, environment management, and traditional DevOps type activities. Platform management looks at uptime and performance monitoring, infrastructure optimization, technology upgrades, cost optimization and future proofing of the IoT ecosystem. This is a much more involved business management activity compared to traditional back-office housekeeping of infrastructure. There is no one single tool or solution provider that can span across managing the entire platform of your IoT ecosystem. Your framework has to account for the tools, processes, and organization to address these platform management needs.

1.1.8 Analytics

With the power of all data, now you need to build the infrastructure for running powerful analytics which actually delivers all the insights. Let us explore the different types of analytics common in the IoT ecosystems.

1. **Massively parallel data analytics**—Massively parallel data analytics or processing (MPP) is an extension from traditional data warehousing. In this technique, similar to MapReduce, multiple processors are employed to run analytical processing. Large data sets are broken up into smaller chunks and worked upon by specifically assigned processors with their own operating system and memory. Messaging interfaces aggregate the results upon completion of analysis by each individual processor. This allows significant performance improvement in the processing and flexible scalability. MPP is applied more on structured data rather than unstructured data. Another drawback of MPP is the need for specialized hardware as opposed to any generic hardware used by other Big Data technologies like MapReduce programming. But MPP is an easy transition for organizations that are making a transition from data warehousing to Big Data.
2. **Time series analysis—**IoT has taken forward the capabilities to collect real-time operational or performance data from devices and machines at very frequent intervals. This gets us the large quantities of streaming time series data for analy-

sis [12, 14]. This type of data is reviewed and analyzed over both time and frequency domain, often to forecast future events and trends. Often multiple parameters from time series data and other environmental or influence data are combined for fault detection and diagnostics (FDD). While the principles and algorithms around FDD have been around for many decades, automated FDD implementation was limited because the lack of availability of adequate data now enabled by IoT and storage/processing of that huge quantity of data enabled by Big Data. There are different types of FDD analysis—rule-based, model-based and case based are popular ones. Rule-based techniques use boundary conditions of observed data to detect faults and isolate them; think of rule-based algorithms as simple or complex if/then/else/hence logic blocks. Model-based techniques use different types of statistical models in more applied analysis. Case-based reasoning uses past situation-solution examples to adapt new solutions. This is how new behaviors are learned over time by modifying analytical parameters.

3. **Machine learning**—Machine learning uses data mining techniques and other learning algorithms to build models of what is happening behind some data so that it can detect, classify and predict future outcomes. These can be used to solve complex problems, predict scenarios previously not modeled or develop new insights. Machine learning helps us here through the use of statistical methods to self-identify algorithms for analysis and association. The popular machine learning algorithms used in IoT ecosystems are association rule learning, artificial neural networks, inductive logic programming, clustering and Bayesian networks. Machine learning has been used extensively in medical diagnostics, predictive maintenance, financial services, and online advertising. Machine learning can be executed either through custom programming or usage of standard or open source tools such as SciKit-learn [28], Spark's machine learning [27].
4. **Advanced data mining**—Data mining is about using statistical data analysis as well as other algorithmic methods to find patterns hidden in the data so that we can explain some phenomenon. Data mining may involve unsupervised knowledge discovery based analysis of very large volumes of data mining is very exploratory in nature, it resides on the crossroads of data sciences, statistics, machine learning and artificial intelligence. Usually, it involves popular classical statistical approaches to regression analysis, neural networks, and decision trees. Sometimes we see analysts applying specific techniques like association rules for initial data exploration, fuzzy data mining approaches, rough set models, support vector machines, and genetic algorithms to solve special problems. Data mining is applied to a wide array of industries to solve a number of complex issues or generate insights which usually involves the interplay of multiple different types/classes of data that may not have obvious associations. Data mining has been applied in the production of salt to the manufacturing of space shuttles; from creating engineering insights to developing insights in the entertainment industry. There are methodologies like Cross-Industry Standard Process for Data Mining (CRISP-DM) and Sample-Explore-Modify-Model-Assess (SEMMA) used in data mining. Often these methodologies are applied in iterative cycles to get the best analysis.

5. **Business intelligence**—Business intelligence (BI) in an IoT environment involves enabling easy creation and access to dashboards and analytics across multiple data sources and types for less technical users. Now we can also create packages which allow one user to distribute her content and analytics to others in the organization or industry who can then conduct their own analytics and create dashboards/reports. BI enables IoT analytics and visualization for masses. There are a number of well-packaged solutions today which have powerful BI capabilities on IoT-enabled Big Data ecosystems; and the output of these can be delivered through multiple channels around the desktop, mobile, and the web.

1.1.9 Security, Identity, and Privacy Management

IoT has amplified the significance of security, privacy, access control and compliance management.

Cybersecurity is one of the biggest issues in front of the technology industry. Every day hackers are inventing new methods to penetrate your technology infrastructure, and companies/technology providers are creating new ways to build barriers against those threat [29–31]. The IT backbone of an organization is not the only pipeline to penetrate; every device that digitally connects to a business or organization adds vulnerabilities. With millions of IoT-enabled devices get-ting connected, the problem is exploding. Moreover, the computing capabilities of IoT devices is very high, which enables malware to independently operate from a device level. With the fragmentation of the technology component providers in the entire IoT ecosystem, there are no natural incentives for any specific player other than the eco-system owner to have end-to-end cyber security responsibilities.

In response, both solution providers and customers of those solutions are implementing multi-layered multi-dimensional intrusion detection and prevention mechanisms. You need security at device level, sometimes data level, definitely at network communication level and usually at the application level. Data and connections need encryption at every stage. Each device and nearly every data transaction need a unique identity which will provide easy traceability and enable authentication. Your security design and policies have to account for perimeter protection, intrusion detection, response and audit trails—all at real-time warp speed. Either you should immediately be able to isolate any non-secure device or connection or have a mechanism to accept them with calculated risks. You should also have mechanisms to detect pattern anomalies in device connectivity, network traffic or data communication; these are instant indicators of a possible intrusion. There are good evolving standards like ISO/IEC 27001 [32], NIST 800-82 v2 [33], RFC2196 [34], ISA/IEC-62443 [35] which you can refer to in designing your policies and practices.

With all the advanced analytics now possible, people and businesses are worried that their behaviors are susceptible to outside scrutiny which might infringe their privacy. The problem gets accentuated because due to geopolitical concerns, national governments and security agencies have taken extreme steps in event and insight detection which has got a lot of bad PR. To increase the value of IoT ecosystems we need more data and more devices to be connected, which works counter to the potential privacy concerns. Anonymizing data coming from devices or processes and insights created from them is a common way of dealing with privacy. This is a delicate balance because too much anonymizing dramatically diminishes the singularity of source insights. You also need to consider policies and practices that do not allow an external intruder the ability to piece together data and insights decomposed into smaller packets to address security and privacy needs.

We need to be careful and transparent with their privacy policy. With IoT, we have a lot of information about our customers. An internal policy has to be established about what we can and cannot do with the information. We need to be transparent with the customers by providing explicit opt-in/out options, and a clear exchange of value. Coupled with privacy comes concerns around who has access to what data and what actions can one take with the data generated. There are many different actors or persons involved—the devices and sensors themselves, manufacturers of these devices and sensors, installers, owners, users, maintainers, application service providers, technology platform providers, cloud hosting providers, etc. Each of these actors has different data access needs and should possess different action rights with the data and insights. There is the added problem of blurred boundaries between many of these actors. On top of that, the roles or persona keep evolving fairly dynamically. Usually, we have never thought of data in terms of its lifecycle, we usually take a static value of data. IoT is changing this archetype, now the context and value of data could be different over periods of time and analysis. Historically there used to be broad classes of groups with well-defined access rights. Now you need to design your access rights policies keeping in mind the dynamic nature of needs/participation evolution and should also be able to define those at a much deeper micro level of data and analytics.

For many years, the IT industry has been exposed to compliance requirements like SOX, HIPAA, GAAP, COBIT, ISO 27002, EUDPD, GLBA, 21 CFR Part 11, etc. [36–38].

These addressed largely financial, often privacy and sometimes security/safety aspects of in-formation. Over the years, the IT applications and organizations have matured to deal with these compliance requirements. IoT brings new considerations to compliance management. First of all, your IoT ecosystem comprises of multiple information ecosystems, so your compliance management design has to address multi-layered and distributed architecture of information. Secondly, the volume, variety, and velocity of data being generated, collected, stored and processed is increasing by many orders of magnitude; this requires your compliance framework to be very adaptive and fast. Finally, since your IoT ecosystem data has to interface with many other data ecosystems like those of your traditional IT systems, the interaction between compliance requirements of these different environments need to be thought through.

One of the key considerations for building your IoT framework is determining how to create and enforce policies around these topics for the multiple actor groups previously discussed. The ability to drive common requirements and ensure standardized implementation of those requirements around security, privacy, access, and compliance gets very complicated; more and so because of the fluidity in the ecosystem. Actors might change, technology might change, devices themselves might change, requirements might change, and vulnerabilities might change. In this dynamically changing ecosystem, you need a single funnel for communicating, implementing and tracking your policies.

Security is absolutely crucial for any complete IoT solution. Security at both the device and network levels is critical to the operation of IoT. Security is about how much one is willing to spend and has been a challenging topic. One can start by applying best practices to every layer of the network stack, e.g. enforce default password change, data encryption, automating patch delivery and protecting interfaces between product and other devices.

We need to determine for each of the items above, a clear strategy and the end goals, and whether one should develop the required methods in-house, outsource or partner.

In addition to the technical considerations, the decision here should be driven by the value created, and how it will impact the business.

The same intelligence that enables devices to perform their tasks must also enable them to recognize and counteract threats. This does not require an evolutionary approach, but rather an evolution of measures that have proven successful in IT networks, adapted to the challenges of IoT and to the constraints of connected devices. At any given point in time, IoT-enabled products must support multiple users and stakeholders including the owner, building operators, tenants, service technicians, OEM, etc. Remote accessibility of connected products requires complex identity management, while we must provide a unified login and product management experience (e.g., Amazon and Google never ask multiple logins to access different services such as Gmail and YouTube).

1.1.10 Delivery

Delivery is related to channel and distribution strategy and includes off-line distribution of IoT-related products, services and on-line distribution of solution for a certain customer and/or a market. This is where your IoT monetization typically takes effect. You should productize your data and analytics and serve them either online or offline to your internal, external or application customers. Often, APIs are provided to app developers to deliver information products. Off-the-shelf data visualization and exploration tools are commonly employed for rapid delivery and form ecosystem. In case you distribute your products electronically, you will have to create standard dimensions similar to software products in terms of licenses management, patch management, user management, entitlements, etc. You will need integration with your institutional ERP and CRM systems to ensure easy sales, support, and business/operational KPI management.

As we briefed, IoT will bring disruptions in customer relationship management that will require innovative product distribution. Effectively designing, developing and administering your delivery components of the IoT framework will be key to greater adoption of your IoT ecosystem.

1.1.11 Business Support

IoT can be impressive, promising convenience, control, and simplicity. However, back office to support these new gadgets, applications, interfaces, and systems aren't nearly as integrated as we want them to be, leaving us wanting more from the products themselves, and the brands that make and sell them. The integration and optimization of devices remain cumbersome and disjointed, leaving a support gap. This disconnectedness is a byproduct of impressive technological innovation without the service and support transformation needed to underpin interrelated applications. For example, if IoT integrates into "one-click buy" or "Pay now" financial transaction services (e.g., paysimple.com, clearent.com, Zuora, Paypal, BluePay, payscape, etc.), then CRM, ERP, and financial payment service integration is a must.

1.2 IoT Partnering Strategy

IoT will eventually be part of every business. The largest reshape for many organizations is going to be adapting to new business models. The goal is to shift away from competing on price but provide customizable products that create opportunities for differentiation and value added services. It broadens value proposition beyond products to include valuable data and enhanced service offerings. You shall focus on matching with customers' business results with your connected product vision to deliver new values to customers.

As we presented in the previous section, a successful implementation of IoT requires the creation of an ecosystem that can enable companies to develop offerings faster, leverage best-in-class technologies, reduce risks and even crowdsource innovation. Companies must consider many different types of partnerships in executing IoT strategy. Few examples are regional, global, channel, industry organization and technology partners.

When you select platform partners, you must consider (1) pricing (2) features (3) partnership model (4) geographical availability (5) support levels and (6) in-house technical proficiency.

1.3 Summary

The technology industry goes through different hype cycles at different points in time. A few years back it was Big Data, today the talk everywhere is about IoT. Many business and technology leaders are still skeptical about the practical utility, impact, and longevity of such hype cycles. But we feel, these are here to stay. IoT is fundamentally changing the way we look at and manage businesses. The words of Jeff Immelt, Chairman and CEO of GE best describes this phenomenon, "If you went to bed last night as an industrial company, you're going to wake up today as a software and analytics company". Technology companies are inundating the market with so many new perspectives and tools that at times it becomes very complex for businesses to contextualize IoT possibilities for their growth; people are increasingly getting lost in bits, bytes, protocols and statistics. This discussion is an attempt to expose you to some of the essential elements of IoT. Your IoT ecosystem design is an important starting point for leveraging the new capabilities to grow your business; the framework is a guide post to facilitate that.

In the following few chapters, we will discuss technology evaluation criteria in selecting each component in the presented framework and partnership strategies in great details.

References

1. (Buyya and Dastjerdi, 2016) Rajkumar Buyya and Amir Vahid Dastjerdi "Internet of Things, Principles and Paradigms", Elsevier, 2016, ISBN: 978-0-12-805395-9
2. (McKinsey 2015) McKinsey, http://www.mckinsey.com/business-functions/digital-mckinsey/our-insights/the-internet-of-things-the-value-of-digitizing-the-physical-world
3. (Deloitte 2015) Deloitte, The Internet of Things Ecosystem, https://www2.deloitte.com/global/en/pages/technology-media-and-telecommunications/articles/internet-of-things-ecosystem.html
4. (Gartner 2015) Gartner Says a thirty-fold increase in internet-connected physical devices by 2020 will significantly alter how the supply chain operates. http://www.gartner.com/newsroom/id/2688717. Accessed 12 June 2015
5. (Poter and Heppelmann, 2014) Michael E. Porter, James E. Heppelmann, "How Smart, Connected Products Are Transforming Competition", HBR Review, Nov. 2014
6. (Ma, et al., 2014) Ada Ma, Youngchoon Park, Sudhi Sinah, Jignesh M. Patel, Self-Conscious Buildings: Transforming Buildings in the Age of the IoT, Johnson Controls Business Strategy Report, Jan, 2014
7. (Rubino, 2016) Sara Cordoba Rubino et al. Meta Products, Retrieved from http://www.meta-products.nl/book/chapter-1/3-what-is-a-metaproduct
8. (Fuqaha, 2015) A. Al-Fuqaha, et al., "Internet of things: A survey on enabling technologies, protocols and applications," IEEE Communications Surveys Tutorials, vol. PP, no. 99
9. (Michael Vögler, 2016) Michael Vögler, et al., "A Scalable Framework for Provisioning Large-Scale IoT Deployments", ACM Transactions on Internet Technology, Volume 16 Issue 2, April 2016

10. (Park et al., 2017) Youngchoon Park, Sudhi Shina, et al., AUTOMATED MONITORING AND SERVICE PROVIDER RECOMMENDATION PLATFORM FOR HVAC EQUIPMENT, United States Patent Application, 1/12/2017, 20170011318
11. (Arm 2015) https://www.arm.com/products/iot-solutions/mbed-iot-device-platform
12. (OPC, 2016) Retrieved from https://opcfoundation.org/
13. (Modbus, 2016) Retrieved from http://www.modbus.org/
14. (BACNET, 2016) Retrieved from http://www.bacnet.org/
15. (Mahmood 2016) Zaigham Mahmood, Connectivity Frameworks for Smart Devices, , Springer, 2016 ISBN: 978-3-319-33122-5
16. (IEEE 1905.1, 2013) IEEE 1905.1-2013, IEEE Standard for a Convergent Digital Home Network for Heterogeneous Technologies," 93 pp., April 12
17. (IEEE 802.15.4, 2011) IEEE 802.15.4-2011, EEE Standard for Local and metropolitan area networks--Part 15.4: Low-Rate Wireless Personal Area Networks (LR-WPANs)," 314 pp., Sept. 5
18. (Z-Wave, 2007) Z-Wave, "Z-Wave Protocol Overview," v. 4, May 2007, https://wiki.ase.tut.fi/courseWiki/images/9/94/SDS10243_2_Z_Wave_Protocol_Overview.pdf
19. (ZigBee, 2008) ZigBee Standards Organization, ZigBee Specification, Document 053474r17, Jan 2008, 604 pp., http://home.deib.polimi.it/cesana/teaching/IoT/papers/ZigBee/ZigBeeSpec.pdf
20. (Strom, 2016) Retrieved from http://storm.apache.org/
21. (kafka, 2016) Retrieved from https://kafka.apache.org/
22. (EventHub, 2016) Retrieved from https://azure.microsoft.com/en-us/services/event-hubs/
23. (Hadoop, 2016) Retrieved from http://hadoop.apache.org/
24. (Sadalage and Fowler 2012) Pramod J. Sadalage and Martin Fowler, NoSQL Distilled: A Brief Guide to the Emerging World of Polyglot Persistence, Pearson Education, 978-0321826626
25. (GraphQL, 2016) Retrieved from http://graphql.org/
26. (OrientDB, 2016) Retrieved from http://orientdb.com/orientdb/
27. (spark, 2016) http://spark.apache.org/
28. (Scikit-learn, 2016) Retrieved from http://scikit-learn.org/stable/
29. (Infineon 2016) Retrieved from http://www.infineon.com/cms/en/applications/smart-card-and-security/internet-of-things-security/
30. (Weimerskirch and Schramm 2010) Retrieved from http://mil-embedded.com/articles/protecting-systems-unauthorized-software-modifi
31. (Wolf et al., 2007) Marko Wolf, André Weimerskirch, and Thomas Wollinger, “State-of-the-Art: Embedding Security in Vehicles,” EURASIP Journal on Embedded Systems, Special Issue on Embedded Systems for Intelligent Vehicles, 2007
32. (ISO 27001, 2016) Retrieved from https://www.iso.org/isoiec-27001-information-security.html
33. (NIST, 2016) Retrieved from http://csrc.nist.gov/publications/PubsSPs.html
34. (RFC2196, 2016) Retrieved from https://www.ietf.org/rfc/rfc2196.txt
35. (ISA99, 2016) Retrieved from https://www.isa.org/isa99/
36. (SOX, 2016) Retrieved from http://www.soxlaw.com/compliance.htm
37. (HIPPA, 2016) Retrieved from https://www.hhs.gov/hipaa/
38. (COBIT, 2016) Retrieved from https://www.isaca.org/knowledge-center/cobit/pages/overview.aspx

Chapter 2
Making Devices Smart

A large number of devices now come with embedded memory, processing power and communicating capabilities helping them become smart. Connected things allow data to be exchanged between the things and its operating ecosystem including manufacturer, operators, users, and other products [1, 2]. The data collected from these products can be then analyzed to make a better decision-making, improve operational efficiencies and enhance performance of the product.

The majority of devices which can be impacted by IoT are still not smart. When the company wants to place the capabilities of being smart into the device, the decision is mainly driven by the following criteria.

Understanding what information, we don't have today and what sensors are needed to gather data that can help us generate value.
Real-time requirement.
The tradeoffs between costs vs. network bandwidth/throughput.
Requirement of high availability and reliability.
The tradeoffs between costs vs. power and memory needed and how much we want the IoT device to be able to do the analytics on its own.

This chapter will delve deep into what kind of computing power and communications capabilities are required in physical devices and how to incorporate them considering their size, functionality, communication and processing power needs, integration into rest of the IoT ecosystem and economic considerations.

We shall cover the following topics in this chapter:

- Defining a smart device
- Common hardware components
- Edge computing
- Making device secure
- Introducing different communication techniques for devices
- Building communication capabilities in devices
- Making devices network with other devices

S.R. Sinha, Y. Park, *Building an Effective IoT Ecosystem for Your Business*,
DOI 10.1007/978-3-319-57391-5_2

2.1 Defining a Smart Device

In an IoT, a smart device refers to an electronic device that can communicate to other devices and cloud services and operates to some extent interactively and autonomously, where its behavior is characterized by loaded software and operating environment. Several notable types of smart devices are smartphones, tablets, smartwatches, smart bands, smart thermostats, connected jet engines and an autonomous vehicle.

Smart devices can be designed to support a variety of form factors, a range of properties about ubiquitous computing and to be used in three main system environments: the physical world, human-centered environments and distributed computing environments [3, 4].

2.2 Common Hardware Components

The world of IoT sensors and hardware is vast and incorporates a variety of components with varying qualities, sizes, and functions. These devices employee low-power and low-cost processor with limited computing resources, and storage.

In this section, we will guide some of the major components commonly used to create smart products. Other than smart gateways, a typical IoT comes with microcontrollers, storage, power supply, I/O interface, sensors, an output device, and connectivity.

2.2.1 Processing Units

Many IoT devices are built with low-cost microcontrollers and ARM-based processors [5, 6]. However, it is worthy to mention microprocessor-based architecture [7], too. MPUs (i.e., central processing unit or microprocessor) use external memory and peripherals to maximize flexibility in system design and performance while and MCUs (Microcontroller) feature integrated on-chip memory and peripherals to minimize footprint, cost and energy consumption. Microprocessors offer more functionality and faster time to market, while microcontrollers provide a smaller, more cost-efficient solution .

Consider Fig. 2.1, a connected door lock may not need a full operating system. Hence, a subset of operating system functionality should be deployed in the MCU-based door lock. It uses existing wireless connectivity to communicate cloud services or mobile devices. Embedded client represents a subset of the full embedded operating system. It exists as APIs and provides basic secure communication stacks, device registration, deceleration or reporting of telemetry data structures, device management, telemetry data reporting, etc.

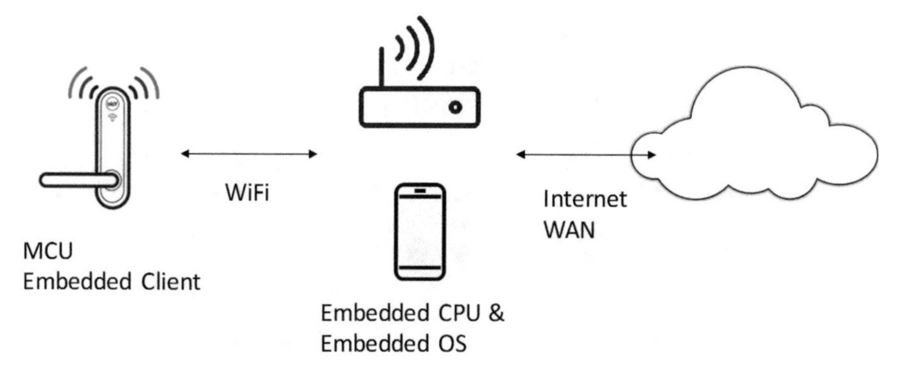

Fig. 2.1 An example of Smart Device Development

MCUs provide faster application start-up. An MCU uses on-chip embedded memory (e.g., flash) in which to store and execute its program. Therefore, MCU has a very short start-up period and can be executing code very quickly. The only practical limitation to using embedded memory is that the total available memory space is finite. Depending on the application, this flash memory size limitation may prove to be a limiting factor. As we discussed, MPUs do not have memory constraints, since they use external storage. Therefore, at start-up software must be loaded into an external DRAM from the non-volatile memory and then commences execution. This means the MPUs will have slower start up time, but the amount of memory you can connect to the processor is relatively larger than MCU based ones. An example of an MCU-based connected device is shown in Fig. 2.2a. It is a reference implementation of the wearable device based on ARM Cortex M3.

The microprocessor can also be a part of IoT solution deployment. As shown in Fig. 2.1, a gateway or router can be x86 processor based. It is connecting the door lock to a remote user with controlled access based on the user's security role. Not only it can be a local data storage to collect and archive data from the sensor/actuator devices, but also it can analyze and compare real-time data against archived historical trends and pre-determined conditions. It can trigger events and initiate appropriate functions to respond to the triggered event, based on a set of pre-configured rules and parameters. Microprocessor-based hardware established a large pool of design resources and a set of specifications adhered to by key companies that deliver x86-based or hardware solutions. Eco-system around X86 CPU [7] includes reference design of hardware system, design guides, developer kits, software components, also include I/O peripherals [8]. Many development resources are available off-the-shelf, enabling the development team to develop and build working proof-of-concept units quickly with minimal cost and effort. These readily available reference implementations and development assistances translate to lower production cost to the final product.

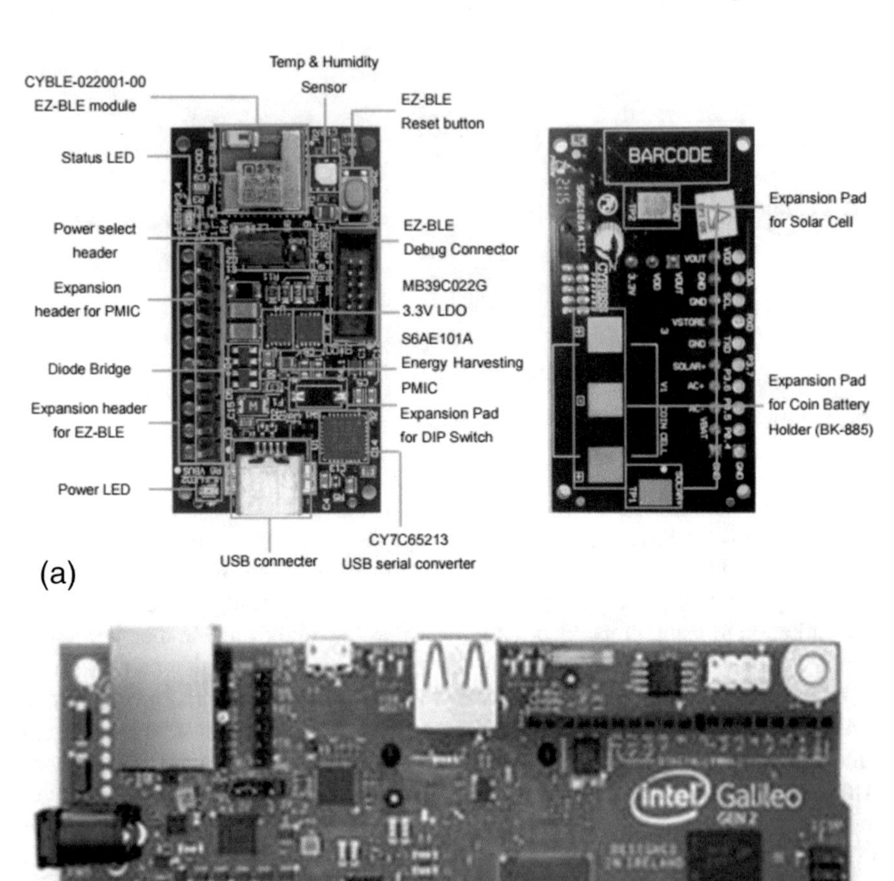

Fig. 2.2 (**a**) Reference design of wearable device from Arrow Electronics based on ARM Cortex M3 RISC, (**b**) Intel's Galileo 2.0 that support full stack operating system [27]

2.2.2 Storage

There are many different types of memory devices are available for use in IoT development. Flash memory is for program space where a program is stored and executed. SRAM is where the program creates and manipulates variables when it runs. EEPROM is a storage space that developers can store long term information. Flash memory devices are high-density, low-cost, non-volatile, fast (to read, but not to write), and electrically reprogrammable. Flash and EEPROM technologies are very similar from software development perspective. The major difference is that flash device can only be erased one sector at a time, not byte-by-byte. Typical sector sizes are in the range 256 bytes to 16 KB. Despite this disadvantage, flash is much more popular than EEPROM and is rapidly displacing many of the ROM devices as well [9].

2.2.3 *Powering Smart Devices*

There are at least three popular ways to powering IoT devices; AC, DC, and battery.

AC is the easiest option. It can provide as much power as a device needs. However, it requires an external AC/DC converter. You could add a small AC/DC circuit onto the IoT device, but this brings safety and regulatory approval issues due to the presence of the line voltage.

DC may be a good option if there is available power source (e.g., vehicle). It is required to have a simple onboard DC to DC converter that can be small, low-cost, and reliable.

The battery is often the most appropriate choice if the battery is properly sized and chosen if it meets business requirements. When selecting a battery, (1) nominal output voltage, (2) energy capacity (3) size, (4) type, (5) source, (6) operating mode, (7) temperature, (8) power drop, and (9) self-discharge rate should be considered.

2.2.4 *Sensors*

Various sensing technologies are available for IoT application development. Examples are vision, position, acceleration, tilt, magnetic, motion, displacement, humidity, leak detection, force, flow, acoustic, vibration etc. [10].

The following list summarizes popular sensing technologies in IoT applications.

The GPS or Global Positioning System tracker detect the global location of the device it is placed on. It receives a signal from satellites and processes them into a position and altitude on the grid of the globe. NMEA message structure is being used exchange data [11].

The camera is one of the most popular sensors that capture images and videos. Accelerometer measures change in motion of an object and can tell other information related to motion such as the direction of movement, tilt, and acceleration of the device it is placed on.

The microphone is to capture and convert sound into an electrical signal.

Gyroscope, often called Coriolis vibratory gyroscope (CVG) uses a vibrating structure to determine the rate of rotation (e.g., an angular velocity or the spin rate of the object it is placed on). Gyroscopes are required in movement analysis such as personal fitness activity tracker or stability tracker in the automotive industry.

Tilt sensor measures a slope in one or more directions of movement.

Piezoelectric sensor measures changes in pressure applied to it. They are usually used for dynamic pressure measurement, such as impact or movement of a body on the sensor. The vibration sensor is an example of Piezoelectric sensors.

Distance measurement sensor detects objects in its line of sight. For example, ultrasonic sensors to determine distance, waves are transmitted in the environment then waves are returning to the origin as echo after striking on the obstacle. LIDAR (Light Detection and Ranging) uses a laser light instead of sound.

The temperature sensor measures the surrounding temperature of the sensor. Most temperature sensors use the resistance changes of a metal or another conductor as a method to detect the temperature.

Humidity sensor measures the percentage of liquid in a target environment by measuring either the resistance, the capacitance, or the conductivity of the material surrounding the probes of the sensor.

Light sensor uses a photocell that converts light energy into electric energy.

Altimeter measures changes in altitude. Barometric pressure sensors are commonly used to measure altitude.

Real-Time Clock (RTC) measures the global time for products that require it.

Radio Frequency Identification (RFID) uses radio waves to identify objects near its antenna. The RFID tags that consist of a microchip and a coil that are responsive to the radio waves sent by the device.

The potentiometer can measure the position of a sliding or rotating knob.

Weight sensor measures weight and consists of one or more load cells and an amplifier.

Often, many different types of sensors are integrated into real-life product development. For example, Fitbit Surge has eight sensors, which include 3-axis accelerometers, a compass, 3-axis gyroscope, altimeter, GPS, heart beat reader, and ambient light sensor [12].

2.2.5 Outputs

Not only sensing devices but also actuating or output devices are required in IoT applications. Often, control or actuation values are written through GPIO (General Purpose Input/Output). Common output devices are a display, a speaker, a motor and a haptic device are important. For example, vibrating motor is a common output device for mobile phones.

2.3 Edge Computing

Cloud computing is not always efficient for data processing when a large volume of data if an amount of data is large and application need real-time actions the edge of the network, and it requires real-time processing [13]. An autonomous vehicle generates about 1-GB data every second, and it requires real-time processing for the vehicle to make correct decisions ("Self-driving cars will create two petabytes of data, what are the big data opportunities for the car industry?") [14]. Other examples are real-time video surveillance camera. It generates a large amount of data and requires real-time intelligent video analytics regardless of the availability of cloud computing. Privacy and security regulations may prevent certain data processing and information sharing in the cloud.

Edge computing is useful when applications need shorter response time, relieve higher network bandwidth requirement, and to meet regulatory and business requirements. It improves time to make a decision, a required action and reduces response time. It while also conserving network resources. The concept of edge computing is not designed to replace cloud computing [15]. However, edge computing can pose significant security, licensing and configuration challenges.

2.3.1 Frameworks for Edge Computing

The basic idea of edge computing is to process data close to the source and leverage locally distributed computing infrastructure. Hybrid computing paradigm has been proposed to provide optimal usage of both edge computing and cloud that may not always be accessible or reasonable. Orsini et al. propose an adaptive Mobile Edge Computing (MEC) programming framework named CloudAware [16], which offloads tasks to edge devices, thus facilitating the development of elastic and scalable edge-based mobile applications. Another example of fog computing framework is Mobile Fog. It provides an API for developing applications which leverage the large-scale, geo-distribution, and low-latency features provided by the fog-computing infrastructure (Fig. 2.3).

Agent-based architecture could be a well-suited framework for edge computing development. Agents are computational entities. It is autonomous and able to cooperate, compete, communicate, act flexibly, and exercise control over their behavior within the frame of their objectives. Cresco [17] is an example of agent-based edge computing framework.

2.4 Making Device Secure

IoT device requires many different types and layers of security along with a set of proper policies around it. Chapter 12 details this security concerns from an end-to-end ecosystem perspective. In this section, we will discuss topics only related to devices.

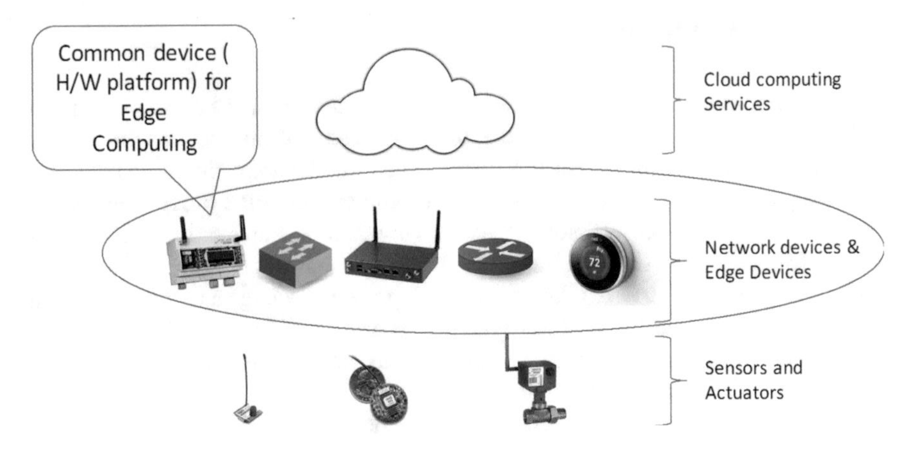

Fig. 2.3 Edge devices and edge computing

- A secure IoT product must provide the following features at a minimum
- Secure Boot
- Secure S/W update
- Data and System Access (system, files, other resources)—Authorization
- Authentication of user
- Secure Communication (from device to cloud, cloud app to app/mobile app)
- Secure Control
- Cyber-attack prevention (stack overflow etc.)—Attack Resiliency
- Device tempering Detection
- Intelligent security policy management (learn and act)

A common set of security features available on the market is based on cryptography (e.g., AES, DES) that provides a secure communication and a digital signature [18, 19]. Hash function offers authentication, integrity, and anti-cloning solution to protect manufacturers and consumers; data and program, that are digitally signed cannot be altered by a malicious third party without being detected by the receiver.

2.4.1 Secure Boot

It is also known as secure, verified or trusted boot. Boot access protection prevents unauthorized booting of IoT devices to stop compromised devices from exchanging data over the Internet of Things [20]. That could be applied to the entire boot process or parts of it. A secure boot scheme uses cryptographic checks to every step of the secure world boot process. This process aims to assert the integrity of all the secure software images that are executed, preventing any unauthorized or maliciously modified software from running [21, 22].

2.4.2 *Secure Software Download*

The secure software download process consists of vital steps are shown in figure below. Once the software is developed, compiled and released, then the object code is passed to a trusted agency that can sign the object code using its private key SK. The signature is then passed back and attached to the program object code. The package of code and signature; usually hash values are immediately stored in a secure database. The appropriate program binary is downloaded to an embedded system. Public verification key (e.g., RAS) is a common method of verification of the validity of the downloaded code [22]. Secure code download process is shown the Fig. 2.4.

2.4.3 *Secure Software Update*

In the software update process, each block of software is encrypted, and the signature of binary code is calculated beforehand. An external programming device or a service authenticates the boot loader. Then the external device passes block by block to the boot loader of the IoT device. The boot loader decrypts and stores each block and computes a hash of it. Once bootloader has computed the hash value over the new flash program file, then digital signature verification start. If the signature verification is successful, the downloaded file is accepted and activated. Otherwise, a safety procedure is activated, and the boot loader awaits the download of a proper flash file [22–24]. As of writing this book, ITU is working on developing standard for updating vehicle software over the air (ITS-AD_08-08).

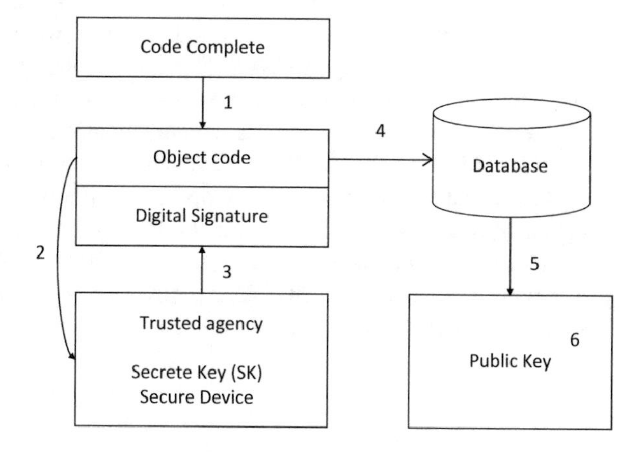

Fig. 2.4 Secure code update and management

2.5 Different Communication Techniques for Devices

There are three distinct types of networks are involved in IoT solution deployments from a device perspective. They are (1) device to device communication (i.e., shown as a sensor and actuator network or an edge device network), (2) device to gateway network, and (3) device to cloud service. Other types of communication protocols and messaging are discussed in Chap. 5.

While, many consumer market (e.g., smart home, personal health) driven IoT devices to connect to cloud services through a WiFi router or a device hub service (e.g., Amazon Alexa or Smart Thing from Samsung), industrial applications require a device to device communication for real-time control and monitoring. For example, a wireless thermostat needs to communicate an air volume control unit to manage the amount of air flow. They communicate through routers or a peer to peer network (e.g., Ad-hoc wireless or wired). When a smart sensor is running on a battery, power consumption of communication device should be a highly important factor. Hence, Zigbee or Z-wave may be the better option compared to WiFi.

Detailed comparisons of various communication technologies are shown in Table 2.1 (Fig. 2.5).

An IoT gateway (see Fig. 2.6) needs to support multiple communication protocols to bridge sensor/actuator network to cloud services. In general, a sensor network protocol and cloud service communication are not homogeneous, hence, a gateway device performs protocol translation (e.g., ZigBee to WiFi) from transportation to application layers. An RFID-based card reader can send cardholder information to a smart door lock controller is shown Fig. 2.6. A smart door lock controller receives a card holder information through RFID protocol adapter and card reader protocol adaptor. Message translation service converts card holder information into a standardized common message that business logic (e.g., decide to open or lock door depends on a cardholder information) can understand and process. A decision of business logic will be translated into a form of a message that WiFi protocol adaptor and door lock protocol adaptor can process, where the WiFi protocol adaptor sends actuation message to WiFi door lock. Door open and close event can be sent to cloud service via connectivity service via 4G LTE.

There are many different criteria we must consider in selecting communication options for IoT devices and gateways. When we deal with end devices such as sensors and actuators, communication distance, commissioning and configuration complexity, battery life, and deployment topology are a minimal set of criteria to be evaluated. When a device requires connecting other devices and cloud services via public wide area network or 3G/4G, the cost to connect and maintenance implication also be a part of evaluation factors, too.

Table 2.1 Comparison of various communication protocols for smart devices

Usage	Protocol	Frequency	Standard	Range	Data rate
Short range wireless – Device to Device – Device to Edge – Device	Bluetooth	2.4 GHz	Bluetooth 4.2 core specification	50–150 m	1 Mbps
	Near Field Communication	13.56 MHz (ISM)	ISO/IEC 18000-3	10 cm	100–240 kbps
	ZigBee	2.4 GHz	IEEE802.15.4 [25]	10–10 m	250 kbps
	WiFi	2.4 GHz 5.8 GHz 5.0 GHz	802.11/a/b/g/n/ac	30 m	1.2–800 MBit/s
	Z Wave	900 MHz	Z-Wave Alliance	30 m	9.6/49/100 kbit/s
			ZAD12837/TTU-T		
			G.9959		
	6LoWPAN (IPv6 over Low-Power Wireless Personal Area Networks)	2.4 GHz	In development	100 m	20/40/250 kbps
Long range wireless Device to Cloud Edge device to Cloud	3G/4G	900/1800/1900/2100 MHz	GSM/GPRS/EDGE (2G), UMTS/HSPA 3G), LTE (4G)	5 km max for GSM 200 km max for HSPA	35–170 kps (GPRS) 120–384 kbps (EDGE) 384 Kbps–2 Mbps (UMTS) 600 kbps–10 Mbps (HSPA) 3–10 Mbps (LTE)
	Sigfox	900 MHz	Sigfox	10–1000 bps	30–50 km (rural)
					3–10 km (City)
	LoRaWAN	867–869 MHz (Europe)	LoRaWAN	2–5 km (urban)	250 bps 5.5 kbps (Europe)
		902–928 MHz (North America)		15 km (suburban area)	980 bps–21.9 kbps (north America)

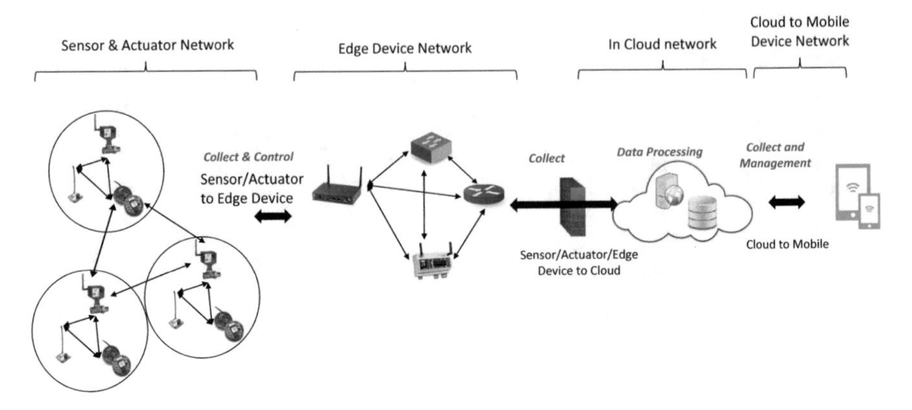

Fig. 2.5 Different communication techniques for connected devices

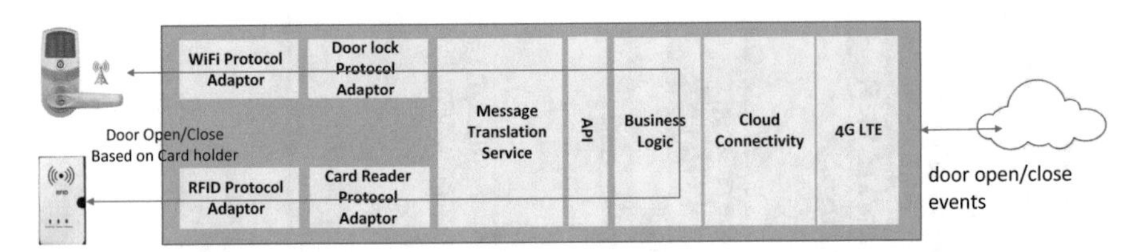

Fig. 2.6 A smart door lock controller

2.6 Building Communication Capabilities in Devices

As shown in the previous section, often we must support multiple protocols in a single IoT device. We could use COTS-based system design to bring communication features into IoT devices. Advantages and disadvantages of COTS-based approach are as follows:

- Advantages:
  - lower development cost
  - ease to evolve product feature
  - the higher maturity of each component
  - hardware and software independent
  - increased interoperability
- Disadvantages:
  - May not be suitable for immediate use
  - May not able to control everything
  - May contain more than what you want

There are many COTS options available from major and minor system chip manufacturers. Most of them offer the system on chip solution that is grouped into three representative categories; high-end, low-end and smart analog device (see Fig. 2.7).

High-end applications can leverage mobile phone architectures. It usually includes processor cores with MMUs, external DRAM, LCD controllers and GPUs. Android or Linux are common operating systems.

Low-end IoT devices are usually microcontroller-based, with embedded flash or lower-cost and more power-efficient non-volatile memory (NVM). A manufacturer's custom real-time operating system common software platform for this SoCs. Often, they include wireless connectivity.

Traditionally, we use an external microcontroller to add smarts to analog devices such as sensors and power management ICs. However, now it is becoming more common to integrate such processor, sensors, non-volatile memory with power management on a die to enable software programmability control things and power consumptions [26].

2.6.1 *Making Devices Network with Other Devices*

Connecting devices through various communication protocols and hubs are common nowadays, however making them interoperable and optimizing the operations between connected devices are not so simple.

Consider Fig. 2.8; current IoT devices are controlled, managed and operated independently from a device to a device. For example, temperature control application does not have any context of operating environment whether a space heater is

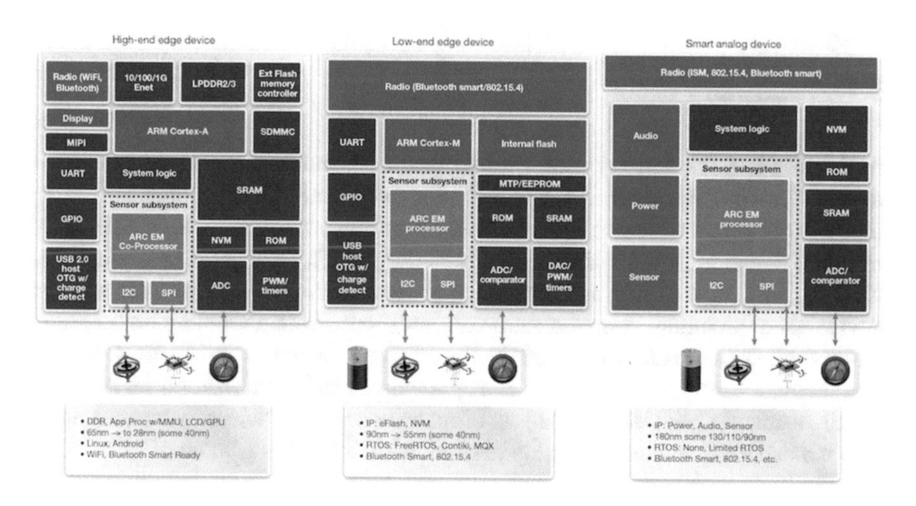

Fig. 2.7 Alternative Architectures for IoT communications (Source Synopsys) [26]

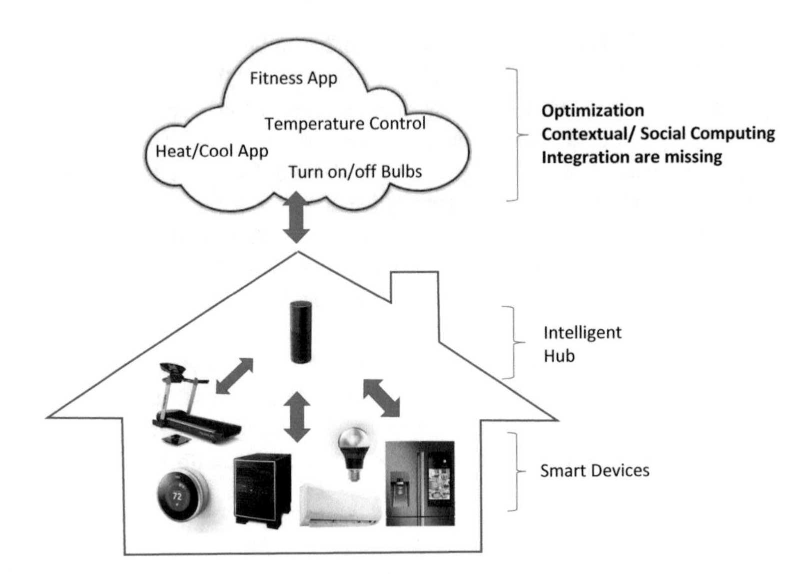

Fig. 2.8 Cloud based optimization of inter-connected devices is missing in the current IoT

on or off, the treadmill is active or not. Simply it will try to maintain a room temperature within an acceptable range. If it is smart and knows the environment, it should suggest turn off a space heater before start cooling. However, current IoT solutions do not fully leverage the integration and optimization that require a different level of semantic or meta-data exchange. IoT needs semantic rich meta-data definition and interoperability standards to make such optimization available.

2.7 Summary

In this chapter, we defined what a smart device is and how to develop such IoT device. Various system components including processors, sensors, actuators, communication solutions, power systems and networking ecosystem. We also discussed relevant communication technologies including WiFi, ZigBee, 6LoPWAN, and NFC, as well as selection and evaluation criteria when we want to build a smart device. At a minimum, communication distance, cost, commissioning and configuration complexity, battery life, and deployment topology must be a part of evaluation metrics. Current IoT solutions and applications lack optimization services that require semantic level interoperability of many connected devices. Advanced metadata management standard is highly desired to unlock the true value of IoT.

References

1. (Poter and Heppelmann, 2014) Michael E. Porter, James E. Heppelmann, "How Smart, Connected Products Are Transforming Competition", HBR Review, Nov. 2014
2. (Ma, et al., 2014) Ada Ma, Youngchoon Park, Sudhi Sinah, Jignesh M. Patel, Self-Conscious Buildings: Transforming Buildings in the Age of the IoT, Johnson Controls Business Strategy Report, Jan, 2014
3. (Mahmood 2016) Zaigham Mahmood, Connectivity Frameworks for Smart Devices, Springer, 2016 ISBN: 978-3-319-33122-5
4. (Greengard, 2015) Samuel Greengard, "The Internet of Things", MIT Press, March 2015, ISBN: 9780262527736
5. (Arm 2015) https://www.arm.com/products/iot-solutions/mbed-iot-device-platform
6. (Cole, 2014) Bernard Cole, "The 8051 MCU: ARM's nemesis on the Internet of Things?", http://www.embedded.com/electronics-blogs/cole-bin/4426602/The-8051-MCU--ARM-s-nemesis-on-the-Internet-of-Things-
7. (Intel 2016) Intel Quark Microcontroller, http://www.intel.com/content/www/us/en/embedded/products/quark/mcu/d2000/overview.html
8. (AAEON, 2016) IoT Gateway Solutions, http://www.aaeon.com/en/c/iot-gateway-solutions
9. (Barr, 2001) Michael Barr., "SRAM or DRAM? EEPROM or flash", retrieved from https://barrgroup.com/Embedded-Systems/How-To/Memory-Types-RAM-ROM-Flash
10. (Lin et al., 2016) Smart Sensors and Systems, Springer, 2015, ISBN-13:9783319147109
11. (Gakstatter 2005) Eric Gakstatter, Retrieved from http://gpsworld.com/what-exactly-is-gps-nmea-data/
12. (FitBit) https://www.fitbit.com/surge#specs
13. (Zicari, 2015), Roberto Zicari, http://www.odbms.org/2015/04/who-invented-big-data-and-why-should-we-care/
14. (Finnegan, 2013) Matthew Finnegan, Boeing 787s to create half a terabyte of data per flight, says Virgin Atlantic, http://www.computerworlduk.com/news/data/boeing-787s-createhalf-terabyte-of-data-per-flight-says-virgin-atlantic-3433595/
15. (WhatIs, n d) browsed from http://blogs.gartner.com/thomas_bittman/2017/03/06/the-edge-will-eat-the-cloud/
16. (Orsini 2016) G. Orsini, et al., CloudAware: A Context-Adaptive Middleware for Mobile Edge and Cloud Computing Applications, Sept. 2016, IEEE 1st International Workshops on Foundations and Applications of Self Systems
17. (Bumgardner 2016) V. K. Cody Bumgardner et al. Cresco: A distributed agent-based edge computing framework, 2016 12th International Conference on Network and Service Management (CNSM)
18. (DHA, 2016) DHS, STRATEGIC PRINCIPLES FOR SECURING THE INTERNET OF THINGS (IoT), https://www.dhs.gov/sites/default/files/publications/Strategic_Principles_for_Securing_the_Internet_of_Things-2016-1115-FINAL....pdf
19. (Abomhara and Køien, 2016) Mohamed Abomhara and Geir M. Køien, "Cyber Security and the Internet of Things: Vulnerabilities, Threats, Intruders and Attacks", Journal of Cyber Security, Vol. 4, 65–88
20. (Infineon 2016) Retrieved from http://www.infineon.com/cms/en/applications/smart-card-and-security/internet-of-things-security/
21. (Li and Xu 2017) Shancang Li, Li Da Xu, Securing the Internet of Things, Elsevier, Jan. 2017, ISBN- 9780128045053
22. (Weimerskirch and Schramm 2010) Retrieved from http://mil-embedded.com/articles/protecting-systems-unauthorized-software-modifi
23. (Wolf et al., 2007) Marko Wolf, André Weimerskirch, and Thomas Wollinger., "State-of-the-Art: Embedding Security in Vehicles," EURASIP Journal on Embedded Systems, Special Issue on Embedded Systems for Intelligent Vehicles, 2007.

24. (Linn 2003) Cullen Linn and Saumya Debray, "Obfuscation of Executable Code to Improve Resistance to Static Disassembly," ACM Conference on Computer and Communications Security (CCS), 2003.
25. (IEEE 802.15.4, 2011) IEEE 802.15.4-2011, EEE Standard for Local and metropolitan area networks--Part 15.4: Low-Rate Wireless Personal Area Networks (LR-WPANs)," 314 pp., Sept. 5
26. (Lowman, 2015) Ron Lowman, A holistic approach to IoT chip design http://www.techdesign-forums.com/practice/technique/iot-internet-of-things-chip-design/
27. (Arrow n.d.) browsed from https://www.arrow.com/en/reference-designs/s6sae101a00sa1002-solar-powered-iot-device-development-kit-provides-an-easy--to-use-platform-for-the-development-of-a-solar-powered-iot-device-with-ble-wireless-connectivity/7069e85c88d7f6f7852a0f2fb1eeabcc

Chapter 3
Creating Smart Gateway

Due to the rapid advent of Internet of Things (IoT), there exists an ever-growing demand for ubiquitous connectivity to integrate multiple data communication protocols, such as Zigbee, Z-Wave, wireless LAN, cable network, etc. The IoT will include not just for modern IoT-enabled devices but also systems that are already in place to-day and operate outside of the cloud-based IoT solutions. For example, RS-485 [1] based Modbus or BACnet are still popular protocols in building automation industry, and those automation devices do not offer full Ethernet or Wi-Fi interfaces.

These legacy devices need an intermediate step or a gateway device to communicate with cloud services that are fast evolving as the cheapest and most capable storage and processing alternatives to on-premise based one. An IoT gateway consists of software and hardware and is to establish and maintain a secure, robust, fault-tolerant connection between the cloud and the edge devices to collect and aggregate device data [2] and to manage the device. It commonly provides the following features:

1. Multiple communication interfaces
2. Network securities
3. Device and connected subsystem provisioning
4. Software management and update
5. Registration of itself and connected subsystems
6. Message routing
7. Message transformation
8. Message aggregation
9. Actuation
10. Protocol translation
11. Access control
12. Web/IP accessible endpoint service of legacy devices

S.R. Sinha, Y. Park, *Building an Effective IoT Ecosystem for Your Business*,
DOI 10.1007/978-3-319-57391-5_3

This chapter will get into details of different types of gateway technology and devices. This chapter will build upon from many of the topics discussed in Chap. 2.

3.1 Hardware Components of a Gateway

Most IoT gateway solutions come with software and hardware. Hardware supports multiple legacy communication interfaces including serial ports, alarm input, and output, etc. It also provides multiple data communication protocols and interfaces such as Ethernet, WiFi, ZigBee, 3G, 4G, Z-wave, 6LoWPAN [3], LoRaWAN, etc. Often, it enables a network bridge between LAN and WAN. Figure 3.1 shows an example of an IoT gateway from AAEON.

Depending on your application, a computing platform needs careful evaluation. There are two popular ones; ARM and x86. The main difference between those in this aspect is ARM instructions operates only on registers with a few instructions (i.e., Reduced Instruction Set Computer) for loading and saving data from/to memory while x86 can operate on direct memory as well (i.e., Complex Instruction set Architecture). Therefore, ARM is a simpler architecture, leading to small silicon area, low cost, and lots of power saving features, while x86 is becoming a power beast regarding both power consumption and production [5, 6].

Intel and AMD's x86 processors are more general and have a relatively larger and established infrastructure for rapid product development. The developer and market ecosystem, and compatibility that x86 has over ARM could be important. ARM designs the CPU architecture and licenses it to other companies including Samsung, NVidia, and Qualcomm, which put their own technologies on top of it and then sell those chips to device makers [7]. This is why Google cannot offer a single OS build of its Android mobile operating system that is based on Linux. Even though, Linux can run on all devices because of the varying hardware configurations. In contrast, with x86, a unified model can be run on various hardware configurations. When you are evaluating gateway hardware platforms, (1) compo-

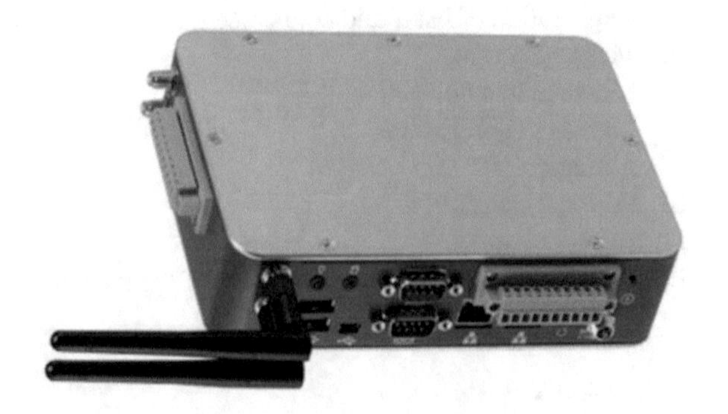

Fig. 3.1 An example of an IoT gateway (courtesy of AAEON Inc.) [4]

nent cost, (2) ecosystem to support development, and (3) development investments, (4) power consumption requirement and (5) maintenance costs must be a part of evaluation metrics.

3.2 Software Components of a Gateway

The software components are the critical element of the gateway. The gateway software is responsible for collecting data from the sensors and other auxiliary devices and storing them appropriately until they can be processed and sent to an application via a target data transmission protocol (see Fig. 3.2). As shown in Fig. 3.2, southbound protocol adaptors are responsible for the discovery of connected subsystems including sensors and actuators and for collecting data. Core services will perform appropriate message aggregation, enrichment, transformation, and transmission, therefore cloud application can consume and processed them.

The gateway software includes intelligent data transmission algorithm to decide if the data at a given stage of processing should be temporary, persistent, or kept in-memory [8]. Intelligent data transmission is required for optimizing data transmission cost when you use cellular network services.

The gateway software should be fault tolerant and must have disaster recovery from failures due to the fact that gateways devices are often operated in the field. Therefore, you should prepare for working conditions that are far from lab environments [9]. For example, the gateway software should be designed for fault recovery from a power outage or a network connection loss that may result in an interruption of intended gate operations. The gateway software should be bootstrapped and started automatically as soon as power returns or network connection restores to the device, and it must continue to work from the point where it was interrupted or failed.

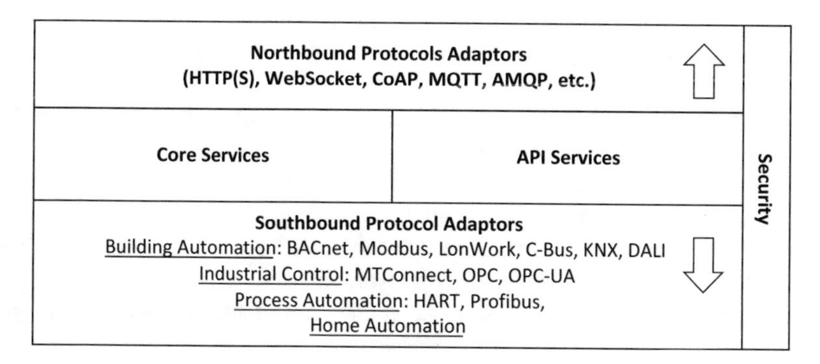

Fig. 3.2 A high-level overview of the gateway software

Gateway software should also be intelligent enough to handle system logging correctly. It must find the right balance between the number of log entries stored on the device and those sent to the data center.

Southbound Protocol Adaptors are designed to connect non-IP and LAN-based IP devices and collect data (most likely pulling data from legacy devices). Plug-in software architecture plays an important role to provide extensibility. Each protocol adaptor shall provide a set of common operations and data models. Therefore, a host process can communicate and exchange data. Recommended common operation interfaces that a protocol driver shall provide via APIs or inter-process communication are as follows:

1. Driver process management including start, stop and restart and kill driver process
2. Driver configuration data management: passing initial configuration data to the driver process, as well as reconfiguration requests, should be available, too
3. Request sub-system discovery
4. Request data reading
5. Request driver performance and status information

A host process that manages protocol drivers are responsible for posting collected data to core services which will optimize the data aggregation, enrichment and transmission at a given stage of processing (e.g., be temporary, persistent, or kept in-memory) and computational constraints.

Northbound Protocol Adaptors are responsible for sending data to cloud services via protocols such as HTTP(S), AMQP and MQTT [10–12].

3.2.1 Core Software Services of a Gateway

Figure 3.3 illustrates component level details of core services presented in the previous section (see Fig. 3.2).

Configuration Web App: this is a web-based application that enables gateway configuration. Common configuration items are network configuration for both WAN and LAN, drivers, users, trending or telemetry data setup, etc. An example is shown in Fig. 3.4.

Gateway command handler: A gateway must provide an interface that a remote user or an application can perform gateway management and unified driver management operations. An operator can update gateway software remotely, can modify or create configuration through a unified gateway and/connected device management console. A management console shall use this command handler interface to manage many connected gateway devices.

Logger: It highly desired to have detailed system logging capability for performance optimization and diagnostics purpose.

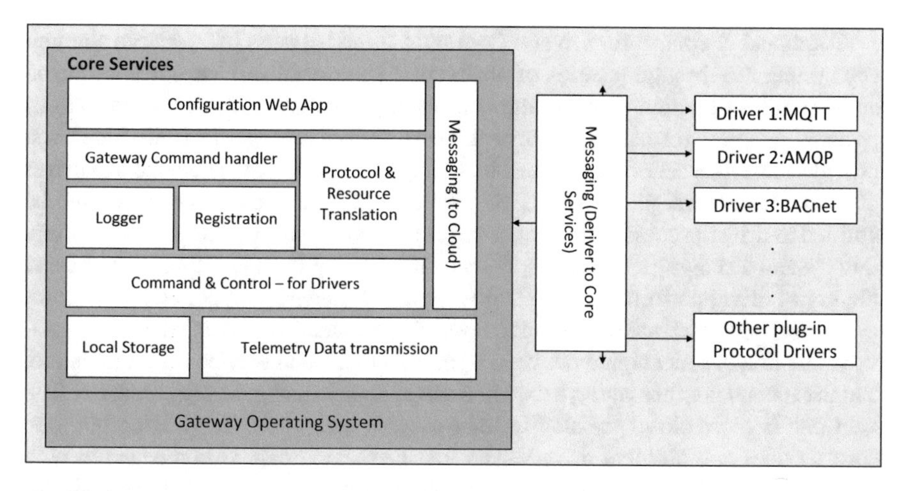

Fig. 3.3 Core services of Gateway

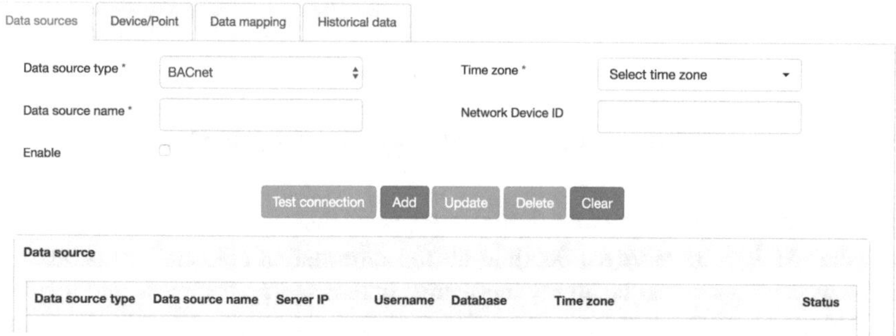

Data sources | Device/Point | Data mapping | Historical data

Data source type * BACnet
Time zone * Select time zone
Data source name *
Network Device ID
Enable

Test connection | Add | Update | Delete | Clear

Data source

Data source type	Data source name	Server IP	Username	Database	Time zone	Status

Fig. 3.4 An example of gateway driver configuration [9]

Registration: A gateway is a connected device, too. It shall be managed as the IoT device that must be registered and provisioned. Detailed registration processes will be discussed in the later section.

Resource & Protocol Translation: legacy data points (i.e., a resource) to be exposed as a trend to the cloud services or to be updated from remote mobile applications are not directly accessible from Internet protocol (e.g., a resource in RESTful protocol), hence, the resource and protocol translation service provides a mechanism to create a virtual resource accessible via IoT service and to perform real-time semantic mediation between a legacy system's resource and a corresponding virtual resource while maintaining uniform semantics. For instance, a temperature reading from a legacy BACnet device cannot directly be accessible from the internet. Resource & Protocol translation service will create a virtual resource (e.g., a RESTful endpoint) and make it available to IoT services and applications. This could be a simple mapping table or complex translation service.

Command & control for drivers: Communication between IoT services and legacy system are through a series of abstraction layers and service. This command and control abstraction provide uniform management capabilities among various protocol adapters (or called protocol drivers). Driver management interface must provide a minimal set of instructions including start, stop, and restart and kill driver process. Driver configuration (e.g., IP address of BACnet device) changes shall be notified to driver process, therefore, a driver can update its operating configuration. Sub-system discovery request, on-demand data reading from the legacy device, and inquiry of driver performance and status are also frequently supported command.

Messaging: There are two distinct messaging mechanisms are involved in a gateway. The first is a messaging between a gateway to a cloud service communication, and the second is a messaging between a driver plug-in and gateway processes. The later one is more closely related to inter-process communication among gateway core services and protocol drivers. The first messaging can be implemented with inter-process communication techniques including pipe, message queue, and shared memory, etc. The second messaging can be implemented with a standard IoT messaging protocols include MQTT or AMQP or HTTP(S). Chapter 4 discuss common IoT protocols in details.

3.3 Types of Gateways and Deployment Topologies

A gateway is not just a pass-through proxy device that forwards data from sensors or other devices to services. Sending all the information collected by devices to backend services would be highly ineffective in terms of performance and network utilization. An IoT gateway will perform certain pre-processing of information close to a data source before they're sent to the cloud service. Examples of preprocessing include message filtering, simple pattern detection, and aggregation.

There are three types of gateways, namely (1) Software only gateway, (2) cloud gateway, and (3) on-prem gateway. Their deployment topology is shown Fig. 3.5.

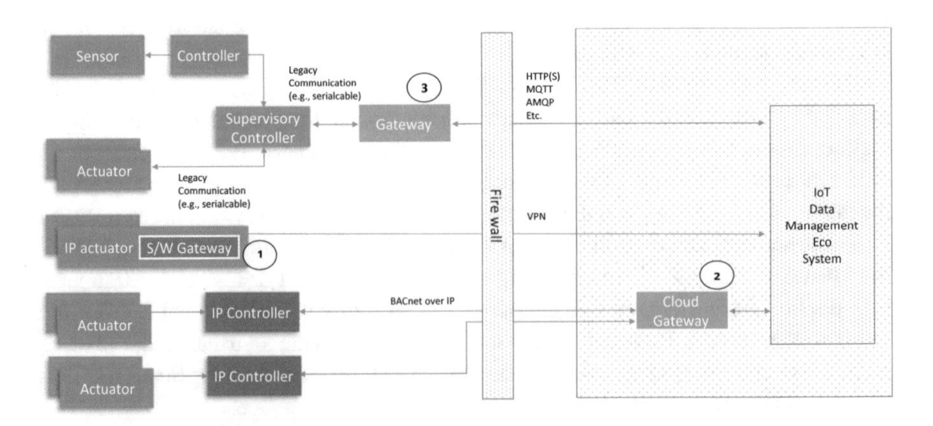

Fig. 3.5 Various types of gateways

Software gateway is usually embedded into an IP-enabled device that uses a software gateway development SDK to make registration, provisioning, and telemetry of the embedded device. SDK comes with common runtimes to make an IP device as an IoT gateway compliant.

Cloud gateway is a software only option that uses legacy IP communication protocols such as BACnet over IP via VPN tunneling. This gateway will perform protocol and message translation between legacy IP protocols and IoT messaging protocols. In addition, it also performs registration and provisioning of a legacy IP device into cloud services.

Finally, the last type of gateway is a physical gateway. It is a physical device dedicated to performing all the necessary feature and functions we have discussed in this chapter.

3.3.1 Software Update of the Gateway

One important feature of IoT gateways is the ability to updates software. After the development of the gateway software onto a device and deliver it to the field, you have very limited capabilities and chances in terms of the gateway software maintenance. The ability to download software updates over-the-air is particularly important from a security and a maintenance perspective, as it can minimize the delivery time of critical security fixes.

The first approach is remote-initiated. In this method, the central software management server pushes the proper version of the software to connected gateways. This requires update push action from a remote management side, shown in Fig. 3.6(1).

The second method is the gateway-initiated method (see Fig. 3.6(2)). With this approach, the gateway is responsible for connecting to the central software repository and compare installed vs. available update. If there is an update, then the gateway downloads the proper version of the software. Software update monitoring agent should be installed in the gateway to communicate with the software management server. This is the most scalable approach because it doesn't require any centralized coordination of the deployment action.

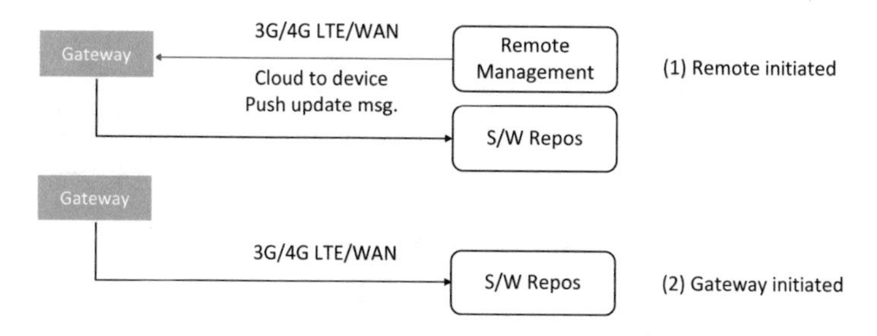

Fig. 3.6 Software update of the gateway

3.3.2 *Registration of a Gateway to Cloud Service*

A gateway should be considered and treated as a smart connected thing. From a device management perspective, it is a just another type of connected thing. It must follow the same registration and provision processes to publish data to cloud services as we do in smart thing registration.

During a manufacturing process of a gateway, security key and unique device identity are loaded and assigned to a gateway. TPM modules are commonly used to keep the secure information.

Once a device is installed, and the network is available, the device will try to connect itself to cloud service. Connecting a device to a cloud service involves two step process namely, registration and claiming.

During the registration phase, a device will obtain security tokens [13] from a security service by using pre-loaded secret keys and an identity during the manufacturing process. Upon successful acquisition of security tokens, the next step is to register the gateway with cloud services (e.g., device registry service). During this registration process, a device shadow is being created and available to other applications. With the correct authentication in placed, security service will register the device into the registry, and it will create primary and secondary keys for future device claim process. As a part of this process, a template of device descriptions including a set of telemetry data point will be registered, too.

At this stage, a gateway is operational (e.g., can send data to cloud service).

Figure 3.7 illustrates the registration and claiming process of the gateway.

A user must claim to get access to the gateway. A user must have an identity that is associated with the unique device identifier. This is called claim process. After completion of claim process, a user may able to subscribe a set of telemetry data points, secure command, and control, or can access device shadow information.

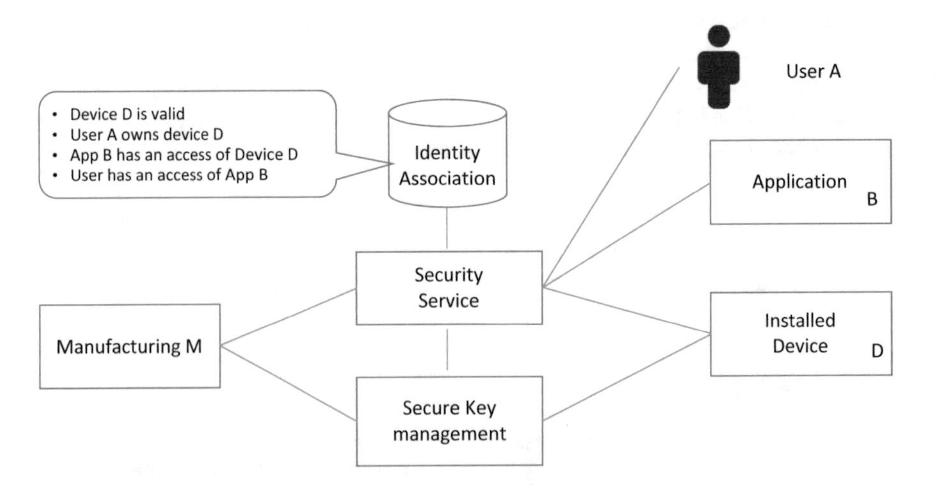

Fig. 3.7 Gateway provisioning and claiming

3.3.2.1 Common Messaging Patterns

Messaging is one of the main functions of an IoT gateway. It collects and sends connected devices' metadata (e.g., device information, etc.,) to cloud services and regularly send telemetry data. It also accepts, translates and distributes incoming messages to correct recipients. In this section, we will discuss common message processing patterns [14].

Message routing: A message routing is a special type of a filter receives messages from a single client or source endpoint and redirects them to the appropriate target endpoint, based on routing rules. It does not modify the message content. For example, an incoming data reading request message from cloud service needs to be routed to correct driver instance (e.g., a message recipient), so it can send back a new sensor reading to the requester.

Message transformation: Message translator modifies the message and translate into a different format. For instance, telemetry data reading from a connected sensor will be translated into a JSON format. Often, we need to enrich the content. The content enricher is a useful pattern when message destination requires more data than is present in the original message. For example, a gateway message enricher can append data source and namespace details into telemetry data.

Message normalization is another popular transformation. It receives heterogeneous message formats from various connected devices and converts them into a canonical message format.

Message aggregation: The message aggregation is a special filter that receives a stream of messages and identifies messages that are correlated (e.g., by time or by location). Once a complete set of messages has been received, this filter collects information from each correlated message and publishes it as a single aggregated message to the output channel for further processing. For example, multiple temperature readings from a connected thermostat over a predefined interval are aggregated and send to achieve service in a cloud.

Message Filtering: If the message content matches the criteria specified by the Message Filter, the message is routed to the output channel (e.g., Cloud service or a specific driver instance). If the message content does not match the criteria, the message is discarded. It is commonly used to eliminate undesired messages from a channel based on a set of criteria. For instance, we may collect may different sensor readings from a connected device, but we are only interested in a subset of readings.

Protocol translation: Often we use more than one protocols in the cloud to gateway communication and a gateway to legacy device communication. A gateway must perform protocol conversion (e.g., message translation with multiple protocol adaptors) to deliver a correctly formatted message to the recipient. For example, an incoming MQTT-based set temperature change message must be translated into BACnet compliant message format to change temperature value.

Web/IP accessible endpoint service of legacy devices: Each legacy data point can be exposed as a resource in REST context. Properly designed namespace (e.g., gateway/driver/data_sample_name/), can be exposed as a resource that can be

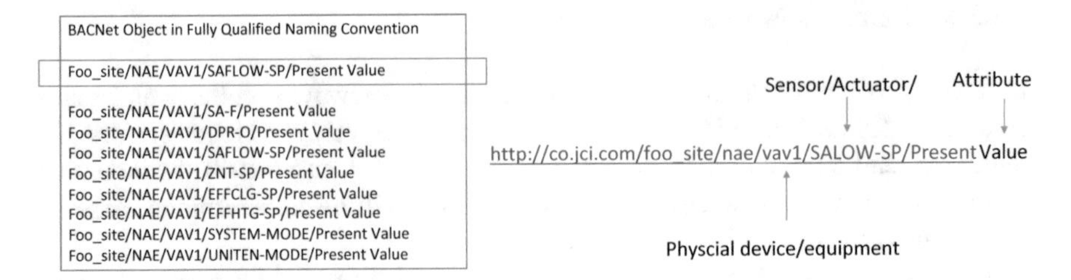

Fig. 3.8 Name translation example to make legacy device to IP compatible

accessed via REST protocol. This makes the legacy device be IP compatible. Figure 3.8 illustrates an example of BACnet point to a REST compliant resource.

3.4 Summary

In this chapter, we have discussed hardware and software components to build an IoT gateway that converts legacy devices into IP compatible. Step by step illustrations of a gateway registration and claiming processes have demonstrated. Two practical methods for remote software update has presented. The first approach is gateway initiated version, and the second is cloud service initiated. Various message patterns have also been discussed from IoT gateway context.

References

1. (Frenzel, 2013) Lou Frenzel, What's The Difference Between The RS-232 And RS-485 Serial Interfaces?, Retrieved from http://www.electronicdesign.com/what-s-difference-between/what-s-difference-between-rs-232-and-rs-485-serial-interfaces, 2013
2. (Sharma, 2014) Sumit Sharma., Retrieved from http://dw.connect.sys-con.com/session/2645/Sumit_Sharma%20.pdf, 2014
3. (Pongle 2015) P. Pongle and G. Chavan, "A survey: Attacks RPL and 6LowPAN in IoT," in International Conference on Pervasive Computing (ICPC 2015), Jan 2015, pp. 1-6., http://ieeexplore.ieee.org/xpl/articleDetails.jsp?arnumber=7087034
4. (AAEON, 2016) IoT Gateway Solutions, http://www.aaeon.com/en/c/iot-gateway-solutions
5. (Lowman, 2015) Ron Lowman, A holistic approach to IoT chip design, http://www.techdesignforums.com/practice/technique/iot-internet-of-things-chip-design/
6. (Cole, 2014) Bernard Cole, "The 8051 MCU: ARM's nemesis on the Internet of Things?", http://www.embedded.com/electronics-blogs/cole-bin/4426602/The-8051-MCU--ARM-s-nemesis-on-the-Internet-of-Things-
7. (LT, 2016) Retrieved from http://www.inetservicescloud.com/linus-torvalds-says-he-prefers-x86-chip-architecture-over-arm/
8. (Konsek, 2015) Henryk Konsek , The Architecture of IoT Gateways, retrieved from https://dzone.com/articles/iot-gateways-and-architecture, 2015

9. (JCI 2015) Design and Implementation of Metasys Energy Gateway, Internal System Design Document
10. (HTTP 1.1, 2014) https://tools.ietf.org/html/rfc7231
11. (AMQP, 2016) Retrieved from https://www.amqp.org/about/what
12. (MQTT 2016) Retrieved from http://mqtt.org/
13. (OAuth, 2012) The OAuth 2.0 Authorization Framework, retrieved from https://tools.ietf.org/html/rfc6749
14. (Gregor Hohpe and Bobby Woolf, 2013) Gregor Hohpe and Bobby Woolf, retrieved from http://www.enterpriseintegrationpatterns.com/patterns/messaging/

Chapter 4
Building Network Services

The messaging and communication technologies discussed in this chapter can be used to connect devices and people (e.g., sensors, mobile devices, single board computers, microcontrollers, desktop computers, local servers, and cloud services) in a distributed network (LAN or WAN) via a range of wired and wireless communication technologies including: Ethernet, Wi-Fi, RFID, NFC, Zigbee, Bluetooth, GSM, GPRS, GPS, 3G, 4G [1–3].

Network services also help with the translation of different messaging protocols between different devices and communication networks. The communications protocols are many and evolving rapidly.

The introduction of IoT relevant transmission protocols (e.g., the sensor transmitting data over the air), including WiFi, Bluetooth, 3G/4G, 802.15.4 and Zigbee. Application layer protocols for various software applications, RESTful API, CoAP, HTTP, AMQP, MQTT will also be discussed too [4–9].

We will also provide a reference architecture for design and implementation of network services in this chapter.

We shall cover the following topics in this chapter:

1. Popular communication protocols
    (a) Translating between different communication protocols
    (b) Understanding different types of communication networks
    (c) Handling messages across various communication networks
2. Handling multi-carrier situation
3. Monitoring and optimizing network performance (e.g., Cost based traffic management)
4. Choosing the optimal communication protocols for your IoT devices and ecosystem
5. Designing optimal communication network architecture for your IoT ecosystem

S.R. Sinha, Y. Park, *Building an Effective IoT Ecosystem for Your Business*,
DOI 10.1007/978-3-319-57391-5_4

4.1 Popular Communication Protocols

We will start grouping of IoT relevant protocols into the following categories; Infrastructure layer: this is corresponding to physical and data link layer in OSI seven-layer model. This layer provides the hardware means for sending and receiving data on a carrier, including defining cables, radio signal spectrum, and physical specification of interfaces including RS-232, Wi-Fi, etc. Data packets are encoded and decoded into bits in this layer.

Management layer: This layer handles switching, flow control, sequencing, routing and error recovery to ensure the completeness of data transmission. It also establishes, manages, and terminates connections between applications.

Application Data layer: this is the layer interacts with software applications. In IoT solution, IoT related services such as discovery, device management, telemetry data handling, and firmware update messaging are relying on application protocols shown in Fig. 4.1. Application set shown in Fig. 4.1 establishes, maintains and ends communication with receiving devices.

All protocols shown in Fig. 4.1 are important. However, we only discuss few physical infrastructure protocols; IEEE 8-2.15.4, ZigBee, and 6LoPWAN, and will spend most of our time to discuss application level protocols; MQTT, CoAP, WebSocket [10], etc.

4.2 Infrastructure Layer Protocols

4.2.1 IEEE 802.15.4

IEEE 802.15.4 [11] is a standard which specifies the physical layer and media access control for low-rate and low-powered wireless personal area networks. The 802.15.4 standard specifies that communication should occur in 5 MHz channels

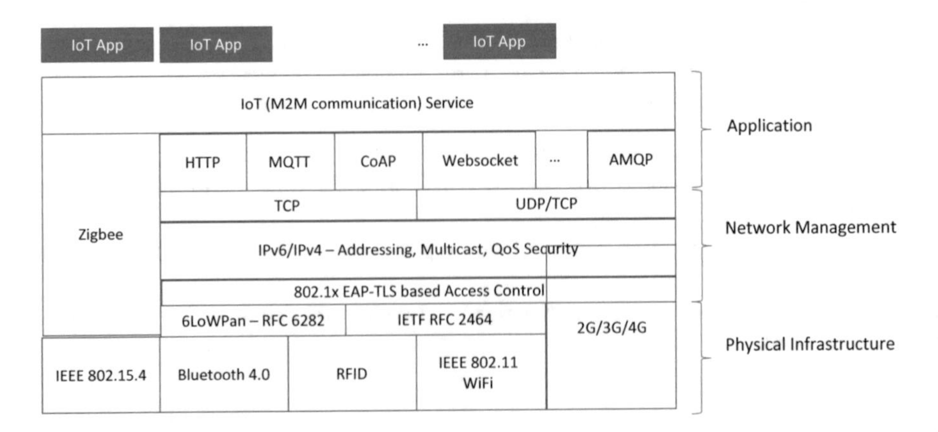

Fig. 4.1 Infrastructure, management and application layer

ranging from 2.405 to 2.480 GHz. Even though the standard specifies 5 MHz channels, in practice, 2 MHz of the channel is used. A maximum over-the-air data transmission rate is 250 kbps in the 2.4 GHz band. However, the protocol overhead limits the actual maximum data rate is approximately half of that. For interference immunity, 802.15.4 specifies the use of Direct Sequence Spread Spectrum (DSSS) [12]. It uses an O-QPSK; Offset Quadrature Phase Shift Keying, with half-sine pulse shaping to modulate the RF carrier. It is the basis for the popular ZigBee specifications. It can be used with 6LoWPAN and standard Internet protocols to build a wireless embedded Internet.

4.2.2 *6LoWPAN*

6LoWPAN [13–15] is a low-power wireless mesh network that every node has its own IPv6 address. Therefore, each node can connect directly to the Internet using open standards. It is an open standard defined in RFC 6268 by the Internet Engineering Task Force (IETF 6Lo). A common network topology of LoWPAN is shown in Fig. 4.2 that includes a 6LoWPAN mesh network. The 6LoWPAN

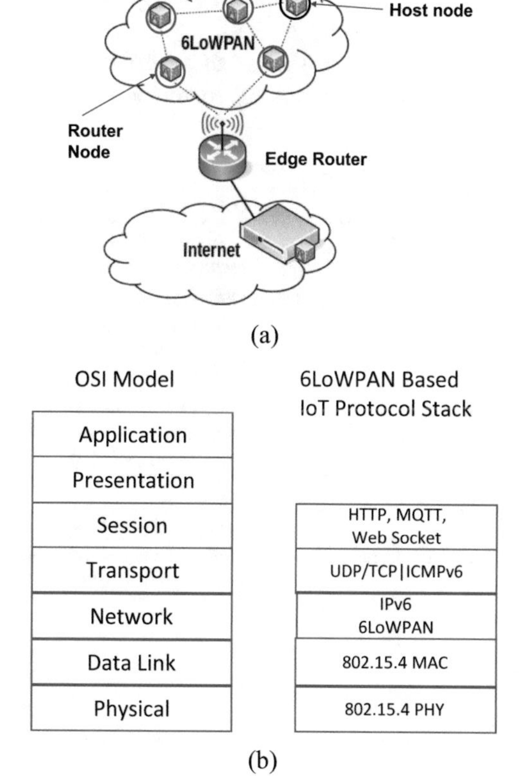

Fig. 4.2 (**a**) A common 6LoWPAN network topology. (**b**) Mapping 6LoPWAN into ISO model

network uses an edge router to connect to the IPv6 network. Often an edge router can connect to other internet routers, too.

There are three types of devices are involved in 6LoPWAN networking. They are the host, router, and edge router.

- The host is an end device that only can send data to adjacent or known router nodes.
- Router routes data to another node in the 6LoWPAN network.
- The edge router provides three functions. First, it exchanges the data between 6LoWPAN devices and the Internet (or another IPv6 network). Second, it routes local data between devices with the 6LoWPAN mesh network. Third, it creates and maintains the 6LoWPAN network (i.e., a radio subnet).

As we briefed, an edge router can connect other IP networks with various options such as Ethernet, Wi-Fi or 3G/4G, since 6LoWPAN only species operation of IPv6 over the IEEE 802.15.4 standard. This is the perhaps the most important distinction of 6LoPWAN compared to other low power wireless protocols. For example, ZigBee® [16], Z-wave [17], or Bluetooth® [8] requires complex application gateways to connect to IP-based networks. These application gateways must understand any application profiles that may be used in the network, and any changes to application protocols on the wireless nodes must also be accompanied by changes on the gateway [18] (Table 4.1).

Figure 4.2b illustrates the mapping between 6LoWPAN-based common IoT stacks and OSI model [19].

Detailed discussion on 6LoWPAN adaption layer, packet header structure, auto configuration, addressing, routing and other specification can be found in [20].

Table 4.1 Comparison between ZigBee vs. 6LoWPAN

Criteria	6LoPWAN	ZigBee
Wireless and IP Interoperability	Provide natural interoperability with other IP network via edge router	Bridging ZigBee and non-ZigBee requires state maintained complex gateway
	Offers interoperability with other wireless 802.15.4 and any other IP network link	
Packet overhead for routing	6LoWPAN links does not necessarily require additional header information	Requires additional header for routing
Security	AES128	AES128
	Per IEEE 802.15.4	Per IEEE 802.15.4
Cost/development cost	Few software development eco-system stacks	Many active open source stacks
		Established ecosystems and many products from various industries including HVAC, Industrial Controls, etc.
IP based	Yes	ZigBee IP—different protocol

4.3 Application Layer Protocols for IoT

In this section, we will discuss various communication protocols to address the following communication situations.

1. Device to Device Communication: message exchanges between devices on a local area network
2. Device to Gateway communication: message exchanges between a device and an internet gateway
3. Device to cloud service communication: message exchanges between a device and a cloud service or a gateway device and a cloud service.
4. Cloud service to cloud service communication: message exchanges between cloud services

Various messaging technologies discussed in this chapter is suited to apply one or more of the communication scenarios illustrated in Fig. 4.3.

4.3.1 Publish and Subscribe Pattern

Publish/subscribe pattern enables easy broadcasting of messages from one publisher to many subscribers. Messages from many publishers to a few subscribers are covered as well. Also, the connection is built-up from the client side, which makes a bidirectional communication possible without NAT translation issues. In the publish/subscribe messaging paradigm, a client, who is called a publisher is sending a particular message from another client (or more clients), who is called subscriber, is receiving the message see Fig. 4.4. This means that the publisher and subscriber don't know about the existence of one another [21]. A broker who is known by both the publisher and subscriber, which filters all incoming messages and distributes

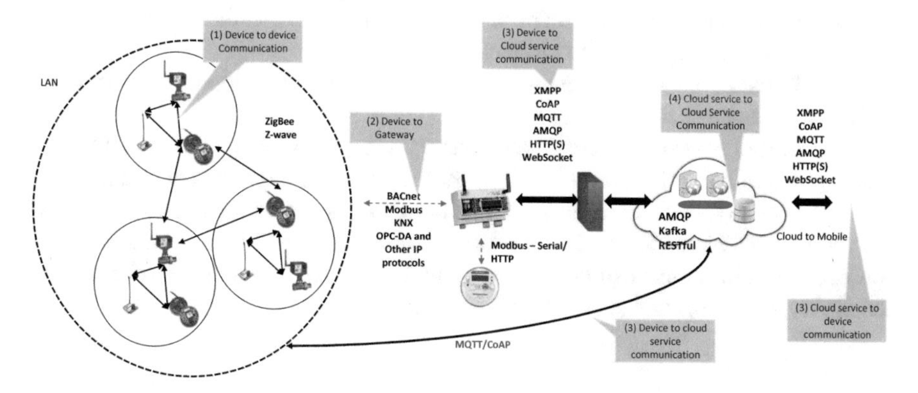

Fig. 4.3 An example of industrial IoT Network Topology to support legacy non-IP devices

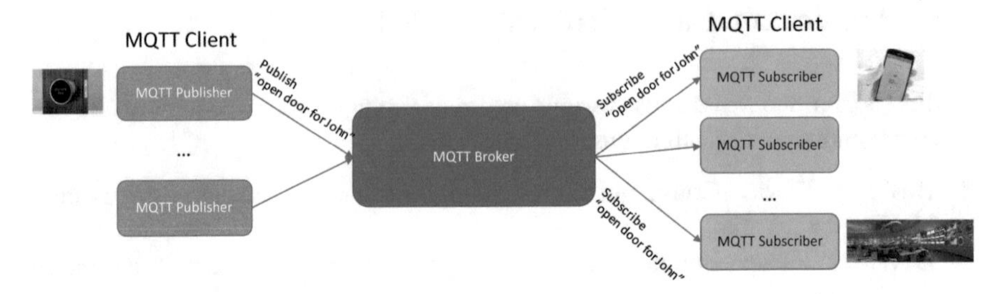

Fig. 4.4 Publish and Subscriber pattern with MQTT

them accordingly. There is three common message filtering types namely subject based, content and type based. For example, MQTT support subject based filtering that are in general strings with a hierarchical structure, which allows filtering based on a limited number of expression. MQTT clients are working asynchronously and are based on callbacks or similar model.

4.3.2 MQTT

MQTT is a low-overhead, simple to implement, ISO standard [22] publish-subscribe-based messaging protocol for use on top of the TCP/IP protocol. It is a message-oriented protocol designed for M2M communications that enable the transfer of telemetry data in the form of messages from devices, along with high latency or constrained networks, to a server or small message broker. It supports publish-and-subscribe style communications and is extremely simple.

Many leading IoT service providers including Amazon, IBM, and Microsoft, support MQTT as a standard device to cloud communication protocol. However, MQTT is constrained to providing primary topic driven messaging in a single namespace, with no long-lived "store-and-forward" queuing pragmatic. This makes it difficult to implement multi-tenant service or to dynamically migrate them or provide simple development to production switch-over.

MQTT uses a password (i.e., a client uses a password to connect Broker) and SSL [23] as main security.

It is worthy to mention few distinctions between message queue and MQTT or other pub/sub based messaging system. In message queue-based communication, message queue keeps messages in storage until they are consumed. Only one client can consume a message (once a message is consumed, it will be removed from the queue). Dynamic queue naming is tough, and the name of the queue must be created explicitly, while topics can be created dynamically.

4.3.3 AMQP (Advanced Message Queuing Protocol)

AMQP [6] is another popular broker based binary IoT messaging protocol that enables conforming client to communicate with conforming message. It was originally intended for the server to server communication to support enterprise messaging scenarios. It is currently supported by OASIS and released as an ISO and IEC standard; ISO/IEC 19464 [13]. It provides various messaging patterns including flow control, round-robin, store-and-forward. AMQP has integration of TLS (e.g., TLS virtual server extensions, known as SNI) and SASL, the IETF set of RFCs that provide appropriate ways of securing the right to use a connection [24]. Furthermore, it is extensible to accommodate modern SASL mechanisms, such as SCRAM-SHA and GS2, and security. AMQP allows separate negotiation of, and policies for, TLS and SASL mechanisms and upwards replacement with alternative techniques as they develop.

AMQP broker consists of exchange, binding, and queue. Exchange receives messages and applies routings. A binding defines rules to bind exchange to queue. Figure 4.5 illustrates a binding example. In general, channel bindings are done via programmatic ways (e.g., channel.binding(exchange="Exchange A", queue = "Queue-1").

There are four different types message routing available in AMQP.

- The direct exchange involves the delivery of messages to queues based on routing keys. Routing keys can be considered as additional data defined to set where a message will go.
- Fan-out exchange completely ignores the routing key and sends any message to all the queues bound to it. This is useful in distribution of a message to multiple clients for purposes (e.g., mass notification), sharing of messages (e.g. chat servers) and updates (e.g. news).
- The topic exchange is for pub/sub (publish-subscribe) patterns. The topic exchange comes in handy to distribute messages accordingly based on keys and patterns.
- Headers exchange constitutes of using additional message headers such as message attributes coupled with messages instead of depending on routing keys for routing to queues. Headers exchange allow differing routing mechanism with more possibilities.

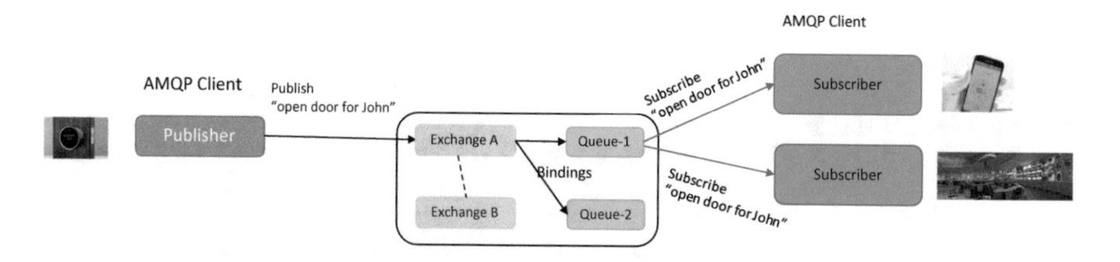

Fig. 4.5 AMQP's exchange, queue and binding

AMQP is a binary messaging system and messages have a payload, which AMQP brokers treat as an opaque byte array. The broker will not inspect or modify the payload. It is possible for messages to contain only attributes and no payload [14]. It is common to use serialization formats like JSON, Thrift, and Protocol Buffers to serialize structured data to publish it as the message payload. AMQP peers typically use the "content-type" and "content-encoding" fields to communicate this information. AMQP messages have QoS properties that include supporting message queuing and delivery semantics covering at-most-once, at-least-once and once-and-only-once (reliable messaging).

Few noticeable open source and commercial version of AMQP are available. Apache Qpid and ActiveMQ are open source version of AMQP 1.0 broker implementation. RabbitMQ and StormMQ are popular commercial version. From IoT perspective, Microsoft's IoT solution support AMQP as Device to cloud service messaging.

4.3.4 CoAP

The Constrained Application Protocol [4] is a specialized web transfer protocol for use with constrained nodes and constrained networks. The protocol is designed for machine-to-machine (M2M) applications such as smart energy and building automation. It is designed for small devices such as sensors. Similar to HTTP, CoAP is based on the wildly successful REST model: Servers create resources and make them available under an URL, and clients access them through standard HTTP calls; GET, PUT, POST, and DELETE.

It has very compact 4-byte header and supports UDP and SMS as transmission protocols (see Fig. 4.6 for detailed message header structure). IP multicasting is available. Therefore, communication between devices is possible (e.g., a mesh-like communication topology). CoAP was designed to minimize message overhead and reduce fragmentation when compared to an HTTP message.

UDP the entire message must fit within a single datagram or a single IEEE 802.15.4 frame when used with 6LoWPAN. Strong Datagram TLS security and built-in discovery are part of a standard. It highly resembles Web/HTTP protocol. A single message must fit into a single IP datagram. Default MTU is 1280 bytes.

Fig. 4.6 Message Header structure of CoAP

0 1 2 3
0 1 2 3 4 5 6 7 8 9 0 1 2 3 4 5 6 7 8 9 0 1 2 3 4 5 6 7 8 9 0 1

Ver	T	OC	Code	MessageID
Token (if any, TKL bytes)...				
Options (if any)...				
Payload (if any)...				

4.4 Choosing the Optimal Communication Protocols for Your IoT Devices and Ecosystem

In the previous section, we discussed various messaging protocols that are summarized in Table 4.2. Selecting a single future-proof communication and messaging protocol for your business is hard. When you are running an IoT-enabled business with paid public internet service, then message encoding and Packet Header overhead should be higher criteria. If your application requires real-time secure and reliable messaging, then AMQP should be your choice. Often, multiple protocol supports are not a bad option if you have a complex business setting. For example, if you are providing smart home control and fire alarm services, then you could choose CoAP for smart home control, but AMQP for a fire alarm [25].

Table 4.2 Comparison of various messaging protocols

	CoAP	MQTT	AMQP	HTTP-REST
Messaging Pattern	Request/reply	Pub/Sub	Pub/Sub plus many others	Request/reply
Packet Header Overhead	Low	Low	Mid	High
Deployment Architecture	P2P	Broker-based	P2P or Broker-based	P2P
QoS	Request/replay—no guarantee	At-most-once, At-least-once, Exactly once	At-most-once, At-least-once, Exactly once	The same as CoAP
Performance/Use	Non-realtime 100s of requests per second	Near-Realtime 1000+ message per broker	Near-Realtime 1000+ message per broker	Non-realtime 100s of requests per second
Security	Datagram TSL, SSL	Username/password	Extensible to external security policies/plug-in	HTTPS/SSL
Message Encoding	Binary	Binary	Binary	Text
Transport	UDP	TCP	TCP	TCP
Recommended	Non-realtime Small device Battery operated	Small device Battery operated	Non-battery powered Need some computing power	Non-battery powered Need some computing power

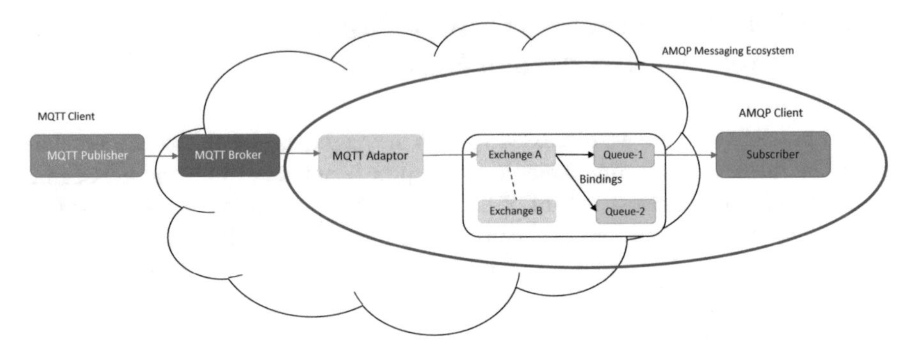

Fig. 4.7 Multi-broker based message exchange

HTTP-REST will be the easiest and fastest way to connect devices over free internet service for OEM.

4.5 Interoperability Among Multiple Protocols

Often, we face integrating other devices or services operating under different protocols. A common way to provide interoperability is through multiple brokers. As shown in Fig. 4.7, we could use MQTT adapter to accept data into AMQP messaging eco-system shown as an ellipse. In this example, MQTT adapter must be able to map MQTT message/topic security into AMQP security model (Routing Key etc.).

4.6 Handling Multi-Carrier Situation

Interfacing multiple telecommunication service providers are common in this connected business environment when connected devices use 3G/4G. Each carrier has own contract terms, conditions, pricing models, unique connected device management interfaces, protocols, security policies, and regulations. These heterogeneities between service providers make a unified connected device management as a complex business and system integration issue in the IoT-enabled business environment. A unified system shall be able to manage the followings;

1. Per device level data service (e.g., manage allowed data usage per billing cycle).
2. Firmware or software update based on schedule or on-demand from a single management console.
3. Integration of network diagnostic/performance monitoring.
4. Integration of billing information from telecom service provider.
5. Automation of invoice processing, and data usage management when multiple carriers got involved in IoT are the foundation for scale out the business (Fig. 4.8).

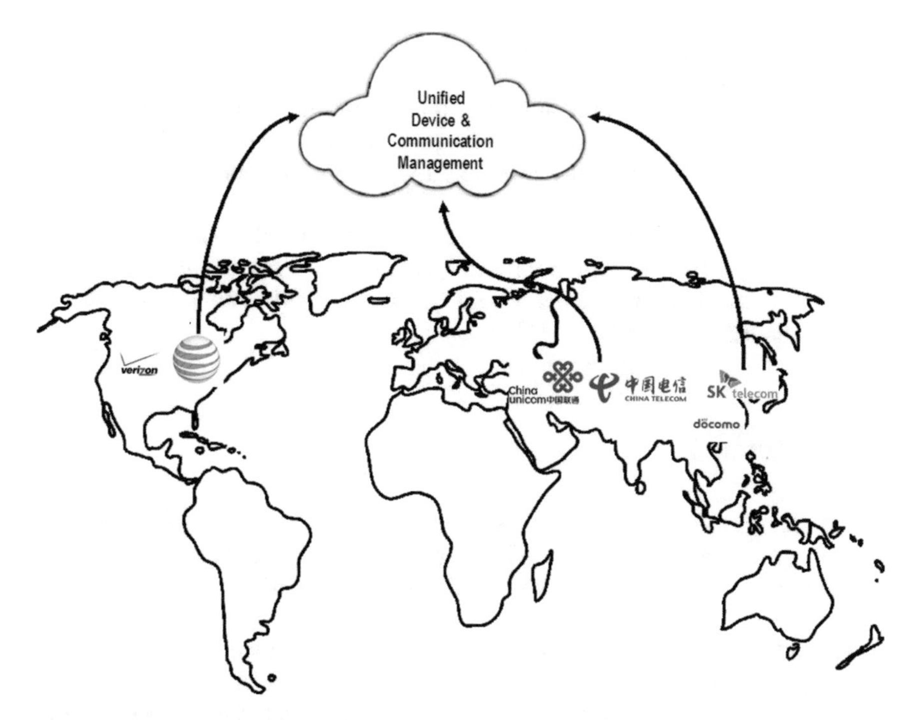

Fig. 4.8 Multi-carrier and international IoT device management

4.7 Monitoring and Optimizing Network Performance

We monitor network performance to improve customer experiences and optimize network usage toward to minimize communication cost associated with. We must able to monitor various network traffic performance. You must monitor all applications that are sending data to a cloud service. Often operating and other applications send, user interactions, logs and other system health information to cloud service that may take a significant position of data usage. Verification of billed data transmission amount vs. actual must be a routine process to avoid billing errors. You should be able to diagnosis and monitor end-to-end, from the application's viewpoint. Some of best practices are

- Chose service and platform provider who can deploy just-in-time network management infrastructure, where needed, when needed, on demand. Establishing mutual customer support with the service provider is critical in operation.
- Employ network and protocol monitoring technologies that provide end-to-end real-time network awareness.
- Once the fundamental networking performance aspects have been addressed, application performance later.

- Use methodologies and technologies that fit your network and needs, not the other way around.
- Continuous monitoring of performance (not just availability) should be an essential starting point. Layer 3 or 4 will be a good demarcation point (at a minimum). Therefore, you can separate network performance issues from application ones quickly.
- Rapid response to network performance problems that slip through the cracks requires (1) real-time measurement, (2) real-time assessment and (3) problem diagnosis capability that delivers quickly, without pre-deploying infrastructure, and should be fixed and resolved remotely.

4.8 Summary

In this chapter, we discussed two protocols for IoT device communication at the infrastructure level, IEEE 802.15.4 and 6LoWPAN, where, 6LoWPAN can be implemented on top of IEEE 802.15.4. We also reviewed three popular messaging protocols; CoAP, MQTT, and AMQP.

For M2M applications. We provide a comprehensive guide for evaluating and choosing a right protocol for your application. The importance of network and application performance with a brief guide to develop monitoring and optimization has presented, too.

References

1. (Gomez 2012) C. Gomez, J. Oller, and J. Paradells, "Overview and evaluation of Bluetooth low energy: An emerging low-power wireless technology," Sensors, vol. 12, no. 9, pp. 11734-11753, 2012., http://www.mdpi.com/1424-8220/12/9/11734
2. (Fuqaha 2015) Al-Fuqaha, A., Guizani, M., Mohammadi, M., Aledhari, M., & Ayyash, M. (2015). Internet of things: A survey on enabling technologies, protocols, and applications. IEEE Communications Surveys & Tutorials, 17(4), 2347-2376.
3. (Dujovne 2014) Dujovne, D., Watteyne, T., Vilajosana, X., & Thubert, P. (2014). 6TiSCH: deterministic IP-enabled industrial internet (of things). IEEE Communications Magazine, 52(12), 36–41.
4. (CoAP 2016) Retrieved from http://coap.technology/
5. (MQTT 2016) Retrieved from http://mqtt.org/
6. (AMQP, 2016) Retrieved from https://www.amqp.org/about/what
7. (HTTP 1.1, 2014) https://tools.ietf.org/html/rfc7231
8. (ISO PRF 20922, 2016) http://www.iso.org/iso/catalogue_detail.htm?csnumber=69466
9. (Pongle 2015) P. Pongle and G. Chavan, "A survey: Attacks RPL and 6LowPAN in IoT," in International Conference on Pervasive Computing (ICPC 2015), Jan 2015, pp. 1–6, http://ieeexplore.ieee.org/xpl/articleDetails.jsp?arnumber=7087034
10. (WebSocket, 2011) https://tools.ietf.org/html/rfc6455
11. (802.15.4-2011) 802.15.4-2011 - IEEE Standard for Local and metropolitan area networks–Part 15.4: Low-Rate Wireless Personal Area Networks (LR-WPANs)

12. (XBee) XBee 802.15.4 Protocol Comparison - digi.com. (n.d.). Retrieved from https://www.digi.com/pdf/xbee-802-15-4-protocol-comparison
13. (OASIS 2012) OASIS, "OASIS Advanced Message Queuing Protocol (AMQP) Version 1.0," 2012., http://docs.oasis-open.org/amqp/core/v1.0/os/amqp-core-complete-v1.0-os.pdf
14. (RabbitMQ, 2016) "RabbitMQ - AMQP 0-9-1 Model Explained" (n.d.). Retrieved from https://www.rabbitmq.com/tutorials/amqp-concepts.html
15. (RFC 6568, 2012) Kim, E., & Kaspar, D. (2012). Design and application spaces for IPv6 over low-power wireless personal area networks (6LoWPANs).
16. (Zigbee 2008) Gislason, D. (2008). ZigBee wireless networking. Newnes.
17. (Z wave 2007) Reinisch, C., Kastner, W., Neugschwandtner, G., & Granzer, W. (2007, June). Wireless technologies in home and building automation. In Industrial Informatics, 2007 5th IEEE International Conference on (Vol. 1, pp. 93-98). IEEE.
18. (Olsson, 2013) Jonas Olsson, 6LoWPAN demystified - TI.com. Retrieved from http://www.ti.com/lit/wp/swry013/swry013.pdf
19. (MS OSI, 2017) "The OSI Model's Seven Layers Defined and Functions Explained" https://support.microsoft.com/en-us/help/103884/the-osi-model-s-seven-layers-defined-and-functions-explained
20. (Hui, Culler 2009) Jonathan Hui, David Culler 2009, Samita Chakrabarti., http://www.ipso-alliance.org/wp-content/media/6lowpan.pdf
21. (HiveMQ 2016) MQTT Essentials Part 2: Publish & Subscribe - HiveMQ. (n.d.). Retrieved from http://www.hivemq.com/blog/mqtt-essentials-part2-publish-subscribe
22. (ISO PRF 20922, 2016) http://www.iso.org/iso/catalogue_detail.htm?csnumber=69466
23. (SANS, 2003) SSL and TLS: A beginners Guide, https://www.sans.org/reading-room/whitepapers/protocols/ssl-tls-beginners-guide-1029
24. (Cohn 2012) Raphael Cohn, https://lists.oasis-open.org/archives/amqp/201202/msg00086/StormMQ_WhitePaper_-_A_Comparison_of_AMQP_and_MQTT.pdf
25. (Granjal, 2015) J. Granjal, E. Monteiro, and J. Sa Silva, "Security for the internet of things: A survey of existing protocols and open research issues," IEEE Communications Surveys Tutorials, vol. 17, no. 3, pp. 1294-1312, 2015., http://ieeexplore.ieee.org/xpl/articleDetails.jsp?arnumber=7005393

Chapter 5
Managing Devices

As the number of internet-connected devices increases, the complexity of management also increases. Hence, we need an appropriate device management solution. Like other products in the market, a connected device has a lifecycle from design, manufacturing, installation, operation or in-use, repair and replacement. In each phase of product lifecycle, there are different requirements from a management solution that shall provide the following to perform a life cycle management of IoT devices.

- Provisioning and Authentication
- Registration
- Configuration Management
- Real-time Monitoring (Central Management)
- Developer support

5.1 Provisioning and Authentication

Provisioning and Authentication [1] are an essential part of establishing a trust-based relationship between a device and a back-end IoT solution and require the upfront development of business processes for unboxing, repair and replacement experience.

Provisioning is the process of enrolling a device into the system or the application [2], Device authentication [3] is part of that process, where only devices with a set of proper credentials are registered. There are many different ways of implementing this process. However, in most cases, the device being deployed is loaded with a certificate or unique identifier that identifies authenticity [4]. Often, TPM [5]; Trust Platform Module is being employed to integrating cryptographic keys into devices during the manufacturing process.

When the device is first connected to the internet, it will start provisioning process. It uses a pre-configured provisioning endpoint and the credentials to obtain authorization grant to access resources in the cloud.

S.R. Sinha, Y. Park, *Building an Effective IoT Ecosystem for Your Business*,
DOI 10.1007/978-3-319-57391-5_5

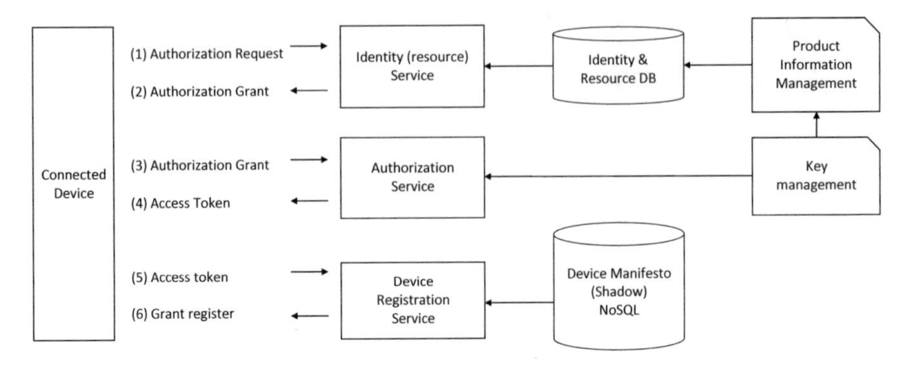

Fig. 5.1 OAuth2 based device provisioning and authentication example

One of popular authorization frameworks is OAuth2 [6] that enables devices to obtain privileges to access resources. An example of OAuth2 based device provisioning and authentication is shown in Fig. 5.1.

As a part of manufacturing process, device identities and credentials (e.g., primary and secondary keys supplied by key management service shown in Fig. 5.1) must be integrated into a local device storage or TPM. Here is a more detailed description of the provisioning and authorization steps.

1. The device will use stored device identity to request permission to register device from identity service that has product identities populated by product information system (e.g., Manufacturing Information Management).
2. If the identity service authorized the request, the connected device receives an authorization grant.
3. The connected device requires an access token from the authorization service by presenting authentication of its identity, and the authorization grant.
4. If the device identity is authenticated and the authorization grant is valid, the authorization service issues an access token to the connected device.
5. The connected device requests the device registration from the device registration service and presents the access token for authentication.
6. If the access token is valid, the device registration service allows registration of a device shadow into device manifesto database.

The actual flow of this process may differ depending applications. However, this is the general idea of device provisioning.

5.2 Registration

In the previous section, we have demonstrated device provisioning process. A logical next step is to register device into a registry to establish an identity for devices and to track device attributes and capabilities. Often document stores (e.g.,

MongoDB, or DocumentDB, etc.,) are employed to persist heterogeneous device description schemas [7, 8].

The registration process is to create a digital representation (i.e., called a device twin, shadow or manifesto), of a physical device into a registry. A registry entry includes attributes and state of a device. A typical set of attributes and state information include hardware specification and installed software details. In general, a separate back-end process in a cloud will maintain consistency between a physical thing and a corresponding digital representation.

5.2.1 Device Manifesto

Device manifesto or device twin is an important concept to serve many device management use cases [9]. For example, a connected device may not able to serve many status information requests from applications, because the connected device may not have enough computing resources to handle or it is disconnected, or it may have unidirectional communication path (i.e., outbound only). In this scenario, a device manifesto provides the device's latest state so that applications can read messages and interact with the device. The device manifesto persisted the last reported state and desired a future state of each device even when the device is offline.

Device manifesto will simplify in developing device management solutions and applications that interact with connected devices by always providing available programming interfaces. Often long-running cloud service will compare the difference between the desired and last reported device state, and command the device to make up the difference.

Common device meta data include a hardware model, an installed software version, device target/commissioned status, reported status, timestamp, unique identifier, meta-data version, and security keys.

Meta-data description schema for end devices such as sensors or actuators is relatively simple compared to gateway device description.

There are two distinct approaches in representing device manifestos, one is a simple device that does not have connected subsystems or devices (e.g., a standalone thing), where another type is a gateway that has many connected systems and subsystems, therefore, we must have a way to model system topology as well as thing itself.

5.2.2 Gateway Registration

Registering a gateway could be multi-step processes including system topology definition. This is very different from a simple device registration processing. First, a gateway uses provisioning steps we discussed to enroll itself. Unlike connected

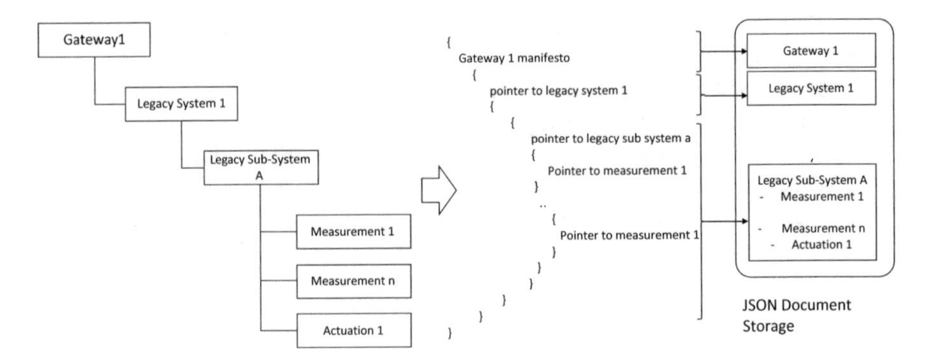

Fig. 5.2 Gateway registration needs different storage and indexing scheme than sensors and actuators

end device such as IP-enabled smart thermostat, the gateway must register attached sub-systems that may include various measurement and actuation points.

As it is shown in Fig. 5.2, a typical gateway is connected to a legacy system 1 via serial communication, where a legacy system 1 controls another legacy subsystem that has multiple sensors and actuators. In practices, describing the entire system shown in Fig. 5.2 as a single JSON document may not be an optimal solution for a fast retrieval and access manifesto (e.g., Hence, subsystem descriptions can be stored as independent JSON documents, and a gateway description will have pointers to subsystem descriptions.

Consider Fig. 5.2; A gateway manifesto may include subsystem descriptions including telemetry data points (e.g., "pointer to measurement 1"). In practice, some connected sensors and actuators could reach tens of thousands of entries that may need different strategies for storage and indexing manifesto. For example, a gateway description may contain subsystem description indices instead of a single large document containing all subsystem descriptions.

5.2.3 *Claiming Device*

Upon completion of a device registration, the next step is to associate applications, users, groups and organizations to a device. Creating relationships (i.e., defining access rights) among the user, owner, organization, and connected device are an essential part of establishing authentication and authorization matrix to service the device and applications. This is known as access and authorization process.

At a minimum, a connected device will be accessed by four district user groups, namely OEM for service under warranty, a service provider for operation and maintenance, users for normal access (e.g., temperature control) and owners. Access model must be carefully designed and administrated provide to prevent potential cyber security issues. Figure 5.3 illustrates these relationships.

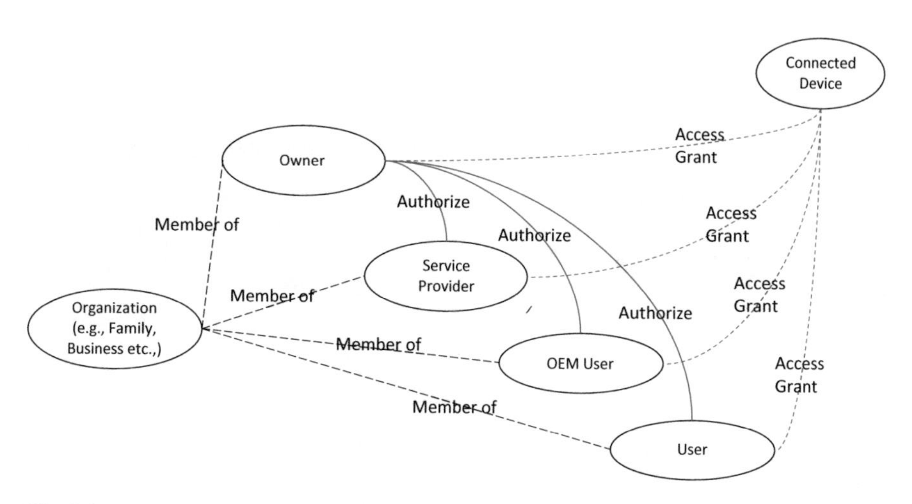

Fig. 5.3 An example of authorization and access relationship among the user, owner, organization, and connected device

An owner can authorize device access to service providers, OEM users, and other users, where federated identity management enables access control simplification [10]. For example, each user will be anticipated by trusted external service agencies such as Google, Facebook or Live account.

Registration service should provide following programmatic and manual interfaces.

1. Register a new device
2. Register a new application
3. Revoke Registration to a device
4. Basic create, read, update and delete on any device
5. Transfer ownership of device from one customer to another
6. Replace the physical device (e.g., in case of failure or upgrade) and restore the existing configuration
7. Managing association between
   (a) a device to a user or
   (b) a device to a user group
   (c) a device to an organization
8. Various querying and retrieval of devices. Examples are
   (a) By location
   (b) By software version
   (c) By user or owner
   (d) By connectivity type
   (e) By status (e.g., connected, connected once, disconnected more than X days, etc.)

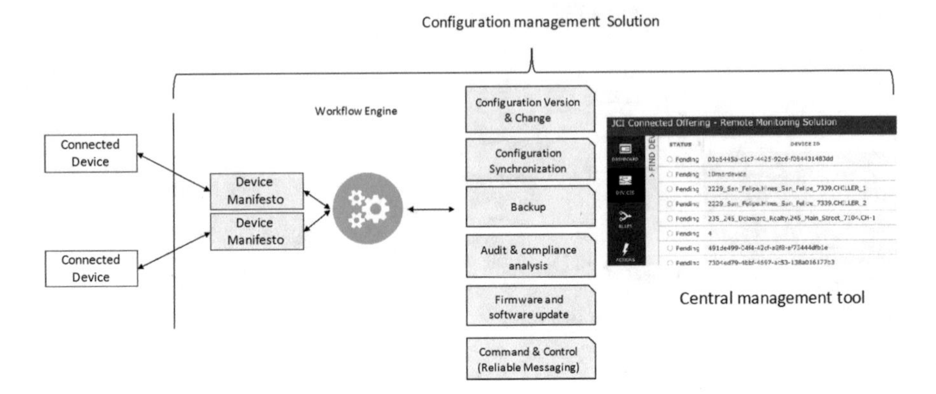

Fig. 5.4 An example of configuration management solution

An application that wants to use a device must go through claiming process to create access and authorization policies among a device, a user/a group or an organization or an application (e.g., a temperature control application for owner John).

5.3 Configuration Management

Configuration management (CM) is related to the process of systematically handling changes to a system in a way that it maintains integrity over time [An Introduction to Configuration Management | DigitalOcean. (n.d.). Retrieved from https://www.digitalocean.com/community/tutorials/an-introduction-to-configuratio]. It requires a set of well-defined management processes and tool chains and services to perform them. An example of a solution is presented in Fig. 5.4.

Configuration version and change management service provide interfaces to create, update, delete and add attributes of connected devices' manifestos.

Configuration synchronization service will make eventually consistency between a manifesto and a physical device.

Connected devices need over-the-air firmware and software updates, but that must be done quickly and securely. Upgrading firmware and operating system are among the most common operations performed by administrators. Firmware and software update help automate these tasks through the use of workflow management, which helps in executing a series of inter-connected commands on a device. All these tasks can be performed on demand or can be scheduled for execution automatically at any future point in time. A workflow engine shall provide simple yet powerful a user interface to provide workflow definitions and optional scripting to program a workflow engine. For instance, a backup service can be considered as a job could be executed via a workflow engine.

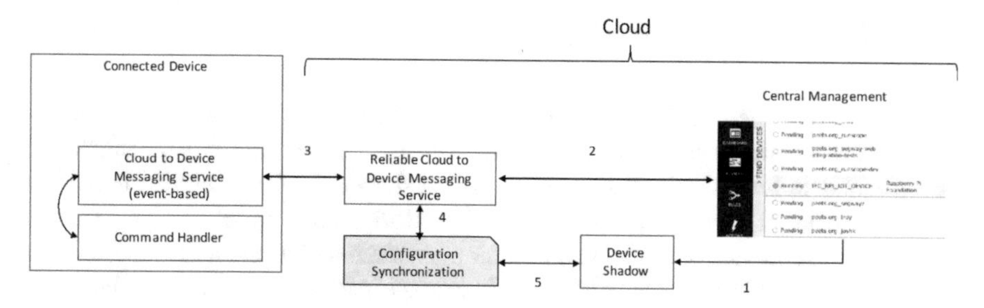

Fig. 5.5 Firmware update scenario over the air

With increasing security concerns to Internet-connected devices and serious legal consequences of information mismanagement and IoT service providers are required not just to meet standard practices and various regulations, but also to demonstrate that the policies are enforced, and connected devices remain compliant to the policies defined. Audit and compliance service shall help administrators to define and enforce standards. Audit and compliance management service will scan the configuration for compliance to the rules or policy defined and report violations.

Configuration management should support various application level communication protocols (see Chap. 3 for details) to provide a reliable messaging between a cloud and an end-device. There are two common messaging patterns. The first is a unidirectional device to cloud data for a device provisioning, a telemetry data transmission, a heartbeat signal reporting, etc. HTTP(S) and MQTT are common protocols of choice to service outgoing messaging.

The second is bi-directional cloud to device messaging to enable command and control. A reliable asynchronous messaging protocol over WebSocket [11] is common to service IoT-enabled device controls from the cloud. Often, an end-device has pre-registered event handling functions that will be triggered by a certain message (e.g., upgrade firmware, increase room temperature, etc.).

Figure 5.5 illustrates a high-level update process flow. First, central management tool tells device manifesto saying that there is a new software update available from a shared repository and second, it sends “firmware update request” message to end device via reliable messaging service. When cloud to device messaging service receives the update request message, it will initiate a firmware update process that will perform upgrade task and update the current firmware version number in metadata section in a device upon successful completion of a software update as well as it will send “successful software update” message to the cloud. This will trigger configuration synchronization task, so configuration of a device and a corresponding device shadow can be synchronized.

Business workflow processing should also be a part of the solution to accommodate various update scenarios and to implement over-the-air system update approval and audit requirements.

5.4 Real-Time Monitoring and Management

Real-time connected device monitoring and management touch various IoT infrastructure components including messaging, security and business rule management. A reference architecture is shown in Fig. 5.6. A device SDK is designed to help to establish secure device connection to an IoT platform and to simplify asynchronous messaging and device status synchronization between a physical and a device twin (usually, long pulling over HTTP(s) or WebSocket from a device is a popular one).

There are multiple communication strategies and methods are required in device management. For instance, turn on a light bulb or reboot a device should be a real-time while availability of new software version notification can be through state synchronization via device twin management.

Change of a value or a device heartbeat or sensor readings (e.g., ON/OFF status of a light bulb) can be delivered through telemetry messaging via message subscription. Often, we use all of three communication methods in device management.

Identity management, access control, and authorization are critical to ensuring that a device can be managed by an application, a group of users and a user with a correctly assigned role and permissions. As we discussed, all device must have a unique identifier in the device registry and assigned security identity via an appropriately managed certificates and key management practices. A device identity must be associated with users, groups, and other applications that will have roles and permissions for device attributes and other resources.

A correctly authenticated device, a user, and an application will be tied to communication channels to establish a role based secure communication environments.

Various business process and orchestration services are required to make a device management solution complete. We could apply "stop software update service" when a device is no longer in warranty period. We could apply schedule updates to a set of devices that are met a sound condition. Device management rules can be easily authored and deployed to an execution engine for schedule.

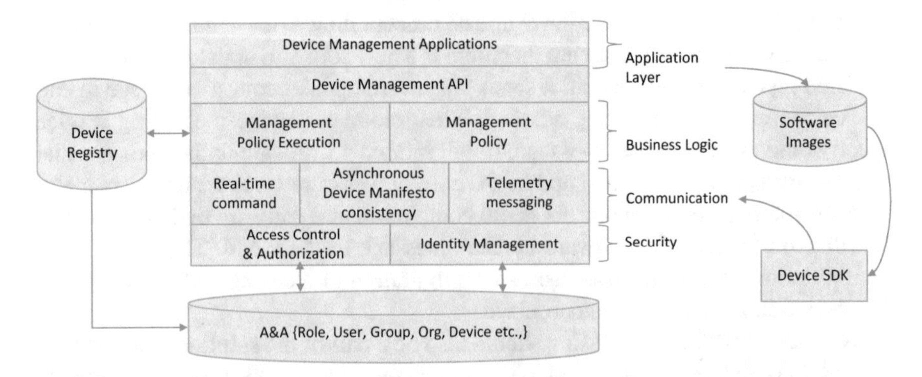

Fig. 5.6 A reference architecture for device management and real-time monitoring application

Many commercial IoT device management solutions encapsulate subsystem elements we discussed in the previous section and make them as a platform as a service, examples are Azure IoT hub [12], and amazon's IoT solution [13].

5.5 Developer Support

Device management shall be a part of customer's IoT solution eco-system that can be integrated into existing operation management solutions including customer support tools, field service management, billing and warranty service, etc.

All management features we have discussed shall be available as APIs. Therefore developers can determine extent and integrate them into other applications. Some of them are

- Security, access control, and authorization API: with these APIs, a developer can obtain security keys and certificates for a device or application as well as associating them with a proper set of permission
- Device Registration: with this APIs, a new device type with various device descriptions (e.g., as tags), attributes, command list and other meta-data can be created in device registry and associated telemetry data entries with returned or reported push/subscription end points.
- Command API: with this APIs, a developer can provide real-time commanding and diagnostics of real-time messaging status.
- Telemetry API: with this APIs, a developer can provide telemetry data push and subscription from a device or a group of devices.
- Application and User Claim API: with this APIs, a developer can make claims of devices that can be associated with users and applications.
- Device information Query APIs: with these APIs, a developer can browse devices based on various filtering options including by owner, software versions, geolocations, date of manufactured, etc.
- It must provide basic create, read, update and delete operations.

Table 5.1 summarizes selection guides for your device management solution.

5.6 Summary

In this chapter, we have discussed an IoT device management that is software running on your fleet of devices and data service platform. With the solution, you can bill, push updates and fixes or command do something for you.

Any IoT system must address device (1) Device Registration (e.g., provisioning and authentication), (2) Configuration and control, (3) Monitoring and diagnostics and (4) Software updates and maintenance.

Table 5.1 Feature requirements for a device management solution

Category	Criteria
Complexity of your infrastructure and devices	Supporting protocols: must be able to communicate multiple protocols and extensible SDK for plug-in-adaptor development Define who will be involved in managing devices, Owner vs. service provider vs. user—complexity of security and user management associated to a device highly depends on this as well as different roles and permission levels
Learning curve	How simple to use and diagnosis
Cost	Total cost of ownership
Development API	Please see developer support section
Support	Can you establish seamless support between your In-house and third party device management support service provider
Flexibility in workflow and job definition	Workflow definition language Workflow creation UI and deployment Flexible scheduling
Required end-point software management	Complexity to update client SDK Client SDK size to fit into small footprint devices
Central management tool	Have user, device, group and application management via single UI and integrated business process management

We demonstrated and showed how we need to two distinct approaches to managing and representing a thing vs. a gateway that are fundamentally different in deployment topologies.

Role-based access control is critical to the IoT device management security and scalability of your solution. With a unified permission, user administration, device registrations with a flexible workflow management, you can control which people on your team, customers can view, change, or execute certain management functions.

Developer support through APIs is a heart of extensibility of the solution. A Minimal set of requirements is proposed and discussed in the later part of this chapter. There are few available [14]. However, commercially available solutions are very limited.

References

1. (Liu L, Yin, 2014) L et al EAC: a framework of authentication property for the IoTs. In: Proceedings of 2014 international conference on cyber-enabled distributed computing and knowledge discovery, Shanghai, China, pp 102–105, 2014
2. (Ssallianz 2017) Retrieved from http://ssallianz.com/remote-management/
3. (Ouaddah, 2017) Aafaf Ouaddah et al., Access control in the Internet of Things: Big challenges and new opportunities, Computer Networks, Volume 112, Jan 2017. Pages 237–262
4. (Sitenkov 2014) D. Sitenkov "Access control in the internet of things". Master's thesis; SICS, 2014

5. (TPM, 2016) Retrieved from https://trustedcomputinggroup.org/work-groups/trusted-platform-module/
6. (OAuth, 2012) The OAuth 2.0 Authorization Framework, retrieved from https://tools.ietf.org/html/rfc6749
7. (DocumentDB, 2016) DocumentDB, https://azure.microsoft.com/en-us/services/cosmos-db/
8. (MongoDB, 2016) MongoDB https://www.tutorialspoint.com/mongodb/mongodb_overview.htm
9. (JCI DM API, 2016) Johnson Controls' Data Platform API, retrieved from https://my.jci.com/sites/PlatformEngineering/DataDocs/Forms/AllItems.aspx
10. (Fremantle, 2014) P. Fremantle P, B. Aziz et al. Federated identity and access management for the internet of things. In: Proceedings of 2014 I.E. international workshop on secure internet of things, Wroclaw, Poland, pp 10–17, 2014
11. (WebSocket, 2011) https://tools.ietf.org/html/rfc6455
12. (MSIOT, 2017) https://docs.microsoft.com/en-us/azure/iot-hub/iot-hub-get-started
13. (AWS, IOT 2017) https://aws.amazon.com/iot/
14. (OMS 2016) OMA LightweightM2M (LwM2M) Object and Resource Registry, http://www.openmobilealliance.org/wp/OMNA/LwM2M/LwM2MRegistry.html

Chapter 6
Performing Data Routing and Real-Time Analysis

As a consequence of the proliferation of connected products and people, we must change our data acquisition and processing strategies to accommodate new data-driven business models and connected product development. A connected product enables us to gather and analyze various data including product usages, updates, operating status, maintenance history and interactions with users while it is in-service. As an OEM, you must able to use collected information to improve and enhance product performances and features. Also, you can develop new offerings based on new insights from data analytics.

Historically data collection from devices, data management, and analysis were sequential steps. Today in the world of IoT, increased volume of data and greater compute capabilities gives us opportunities to quickly impact device level or micro-operational level business process decision making, thereby breaking the linearity from data to analysis. Right after the creation of data from devices or other intermediate processes, we do two things simultaneously—distribute the message to appropriate containers of the IoT ecosystem, and perform complex event processing for real-time analysis to get quick insights leading to quick decision influences. In this chapter, we shall explore both of these simultaneous events. We shall cover the following topics in this chapter:

1. Data acquisition strategy for connected products
2. Introduction to stream processing with IoT relevant use cases
3. System design pattern for stream processing
    (a) Lambda vs. Kappa Architecture
    (b) Stream Processing Systems
    (c) Rule-based complex event processing
4. Common techniques for identifying meaningful events and rules
5. Integration with machine learning

S.R. Sinha, Y. Park, *Building an Effective IoT Ecosystem for Your Business*,
DOI 10.1007/978-3-319-57391-5_6

6.1 Data Acquisition Strategy for Connected Products

Traditional data models, schema designs and acquisition strategies assumed communication and storage cost are expensive. That is now negligible in many cases. The popularity of connected products and market competitions to deliver data-driven innovations make us rethink how we design and capture data from an IoT-enabled product.

A connected product design must take account of data acquisition strategies that includes:

1. What are the new values we can create and offer from new data set?
2. What data must be acquired from a product, an owner, and a user to create values discussed in
3. What are other data from external data services (e.g., data.gov, Dun, and Bradstreet, etc.) needed?
4. Implication to integrity, security, and privacy
5. How often we need to collect them?
6. How long need to be retained?
7. What are the related acquisition costs (e.g., sensors)?
8. Who owns and has usage rights on data?

Data collected from connected products, users, and other development activities are essential to create business values and competitive advantages. We must plan for creating values from data. If you are planning to be in after-market service industry, you must have near-real time telemetry data from connected products and field service teams. Not only from your product generated data, but you will need other data types such as weather, oil price, etc., For example, a smart thermostat will use outdoor air temperature and peak energy demand to optimize home energy usage. A smart connected car service will analyze real-time traffic patterns, and weather conditions and other driver information to forecast a near-real-time risk associated with a customer.

Consider Fig. 6.1a; it illustrates a simple data logging approach commonly used in embedded product development to minimize storage usage. It is a good strategy for stand-alone and non-connected device point of view. However, it is very difficult to understand how a product is being used since there is no operating and user interaction related data available. In contrast, Fig. 6.1b presents an aggressive event logging approach that captures every change made to a connected product. With this rich data, we could infer product design faults if there are ones, or improve product performance.

In the following few sections, we will discuss various technical components to collect, process, and analyze real-time data from connected products and users.

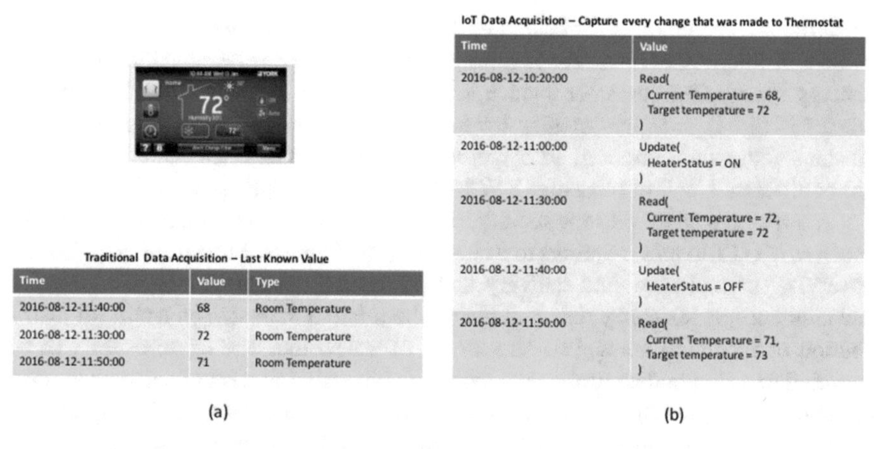

IoT Data Acquisition – Capture every change that was made to Thermostat

Time	Value
2016-08-12-10:20:00	Read(Current Temperature = 68, Target temperature = 72)
2016-08-12-11:00:00	Update(HeaterStatus = ON)
2016-08-12-11:30:00	Read(Current Temperature = 72, Target temperature = 72)
2016-08-12-11:40:00	Update(HeaterStatus = OFF)
2016-08-12-11:50:00	Read(Current Temperature = 71, Target temperature = 73)

Traditional Data Acquisition – Last Known Value

Time	Value	Type
2016-08-12-11:40:00	68	Room Temperature
2016-08-12-11:30:00	72	Room Temperature
2016-08-12-11:50:00	71	Room Temperature

Fig. 6.1 (**a**) Traditional data acquisition vs. (**b**) IoT-aware data acquisition

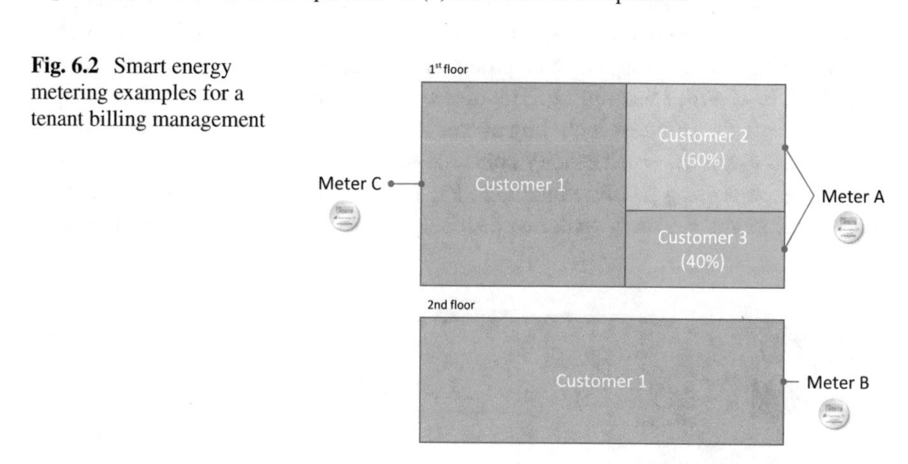

Fig. 6.2 Smart energy metering examples for a tenant billing management

6.2 Introduction to Stream Processing with IoT Relevant Use Cases

Events are everywhere. We have seen various forms of events used in different areas such as event-driven programming such as system notifications in infrastructure and operations, user, message-oriented middleware, pub/sub message queues RFID readings, e-mail messages, and instant messaging, and so on [1–3].

In this section, we will consider energy data management and analytics as an IoT relevant example that requires energy meter data gathering and analysis with enterprise master data such as space or energy zone hierarchy. Smart meters popular in managing and understanding utility and energy consumptions and tenant billing management.

Figure 6.2 illustrates a typical sub-metering setting to charge proper electrical bills to tenants in a building. As shown, an electrical meter A measures two tenants' electricity consumptions for customer 2 and 3, where customer 2 takes 60% of the total consumption of meter A's reading, while customer 3 is responsible for 40% of energy spend. Our customer 1 occupies two separate spaces that are metered by meter B and C. Imagine, meters are Internet connected and send telemetry (e.g., instantaneous consumption) at a pre-configured interval, where each meter may have different a unit of measure (KW/h vs. W/h) and sampling intervals.

A smart meter may use a standard device to a cloud communication protocol such as MQTT to push readings to a cloud service. Figure 6.3 illustrates a high-level overview of real-time data delivery and processing system. An MQTT protocol adaptor accepts telemetry data and places them into a messaging system for distribution of incoming messages. The arrival of a new message or an event will be notified to a stream data processing system that consists of a set of services to meet business and operating requirements (e.g., pre-processing of data and analytics).

Figure 6.4 illustrate one way to implement such streaming data processing system. Incoming data stream are stored as a message on high throughput reliable, distributed, fault-tolerant replicated messaging systems (e.g., Kafka [4] or Eventhub [5]). Each data stream may have its processing requirement such as unit conversion from W/h to KW/h, re-sampling. Therefore, the relevant processing definitions, as well as stream data context including content tags, meta-data, etc., are fetched in real time. As an example, an electricity consumption of customer 1's may be computed with three processing pipeline operators P1, P2 and P3. They are a data cleaner, an add operator, and a stream writer respectively. Data cleanser may perform missing

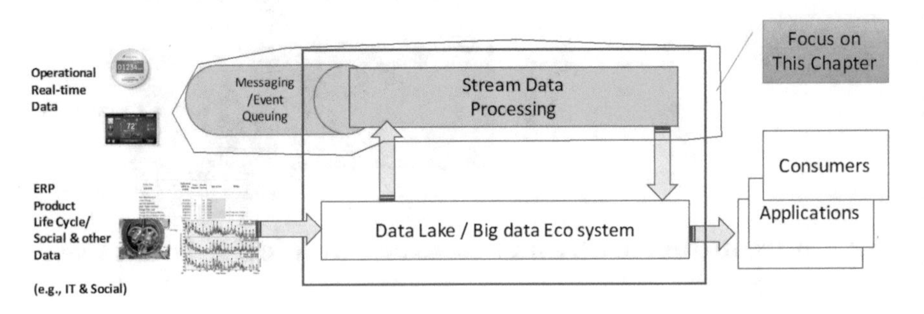

Fig. 6.3 A high-level view of real-time stream processing system

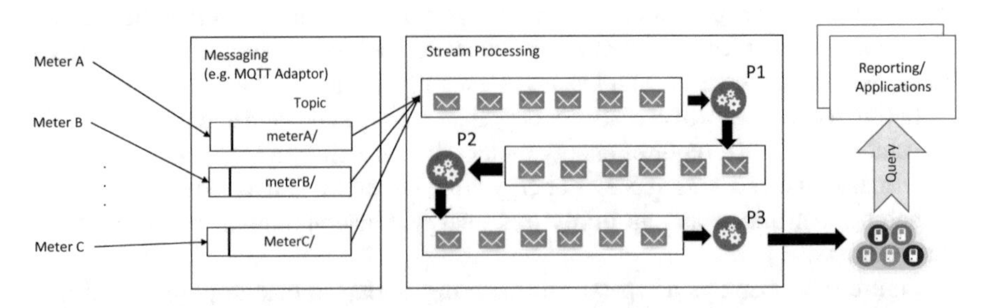

Fig. 6.4 A stream processing example in our energy management example

data handling, outlier removal, re-sampling, unit of measurement conversion, etc. Stream processing execution engine will manage all.

The add operator performs addition of two data streams; namely meterC and meterB and it produces a new data stream that will be input to a stream writer to persist newly computed stream data.

As we briefly discussed, there are many different stream data processing challenges in industrial IoT applications including the smart metering example we are discussing. Network failures, missing data, arrival with out of order, late arrival and noise are common data preprocessing challenges in real life application development, where we must have robust moving window (or called sliding window) operations when we implement stream processing system [6].

Typical stream analytics systems work well when events are arriving into the system in real-time. In other words, the lag between the "event creation time" and "Ingestion Time" is insignificant. As a result, processing windows are typically based on Ingestion time. Often stream processing assumes events arrive in an ordered manner and events don't get updated. Figure 6.5 illustrates data challenges in stream processing.

Unfortunately, in the real-world, when it comes to streaming data from real-world IoT devices, many of these assumptions may not hold true:

- Events may arrive into the system with a lag ranging from minutes to hours to days.
- The lag for a specific device is not predictable (fixed).
  - A device may lag by hours at certain times while it may be fully caught up at other times.
- The lags for different devices may be different.
- The system may be dealing with million different devices with lags ranging from sub- second to several hours to few days.

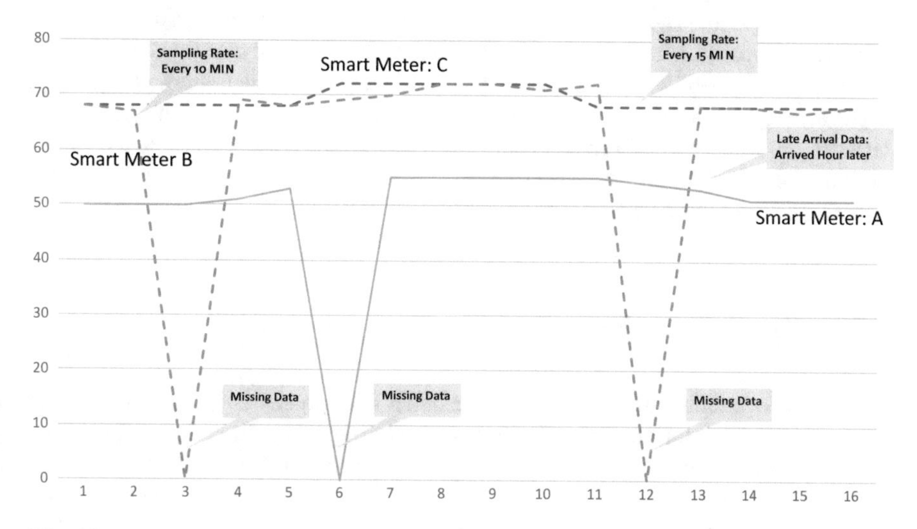

Fig. 6.5 Data processing challenges in industrial IoT

- Events may get "updated."
  - A device may send "incorrect" events for a few days. After this is detected and corrected, the device may send the updated (correct) historical events.
- Events may arrive out of order.
  - System may get an event for time 10.00 AM followed by a burst of events for time 9.30 to 9.59 AM) for a specific device.

As we detailed, stream processing based on "Ingestion Time" will not be sufficient. Instead, we need operations based on "Event Creation Time." Event creation time to be treated as a first-class citizen in stream processing systems for all processing operations. We need a way to efficiently store and retrieve historical data that belongs to a window that needs to be reprocessed because of new data/updated data for that window.

6.3 System Design Pattern for Stream Processing

In this section, we will review two popular design patterns when it comes to implementing stream processing. They are known as lambda architecture and Kappa architecture. At a high level, a stream processing system comprises of (1) time series or events storage and retrieval service, (2) messaging subsystem, and (3) execution management.

With the advent of distributed key-value or document storage, HBbase, DynamoDB, DocumentDB, or MongoDB are popular options for event or time series storage.

A messaging subsystem is responsible for transferring data from one processor to another, so the processor can focus on information processing, but not worry about how to share and distribute it. Reliable message queuing technology is a foundation of distributed messaging, in which messages are queued asynchronously between client applications and messaging system. Microsoft's EventHub or Apache Kafka or Amazon Kinesis or Google Cloud Pub/Sub are popular choice to manage data flows [7–9].

Publish-subscribe system is the most common in stream processing, where messages are kept on a topic. Depends on a messaging technology, topics can be persisted. Consumers can subscribe to one or more topic and consume all the messages in that topic. In the Publish-Subscribe system, message producers are called publishers, and message consumers are called subscribers.

6.3.1 *Lambda vs. Kappa Architecture*

The Lambda architecture comprises of a batch data processing, a real-time data processing, and a serving layer (shown as reporting and applications). Both the batch and realtime processing receive a copy of data stream, in parallel (two distinct paths). The serving layer then aggregates and merges processing results from both layers into a complete answer (see Fig. 6.6).

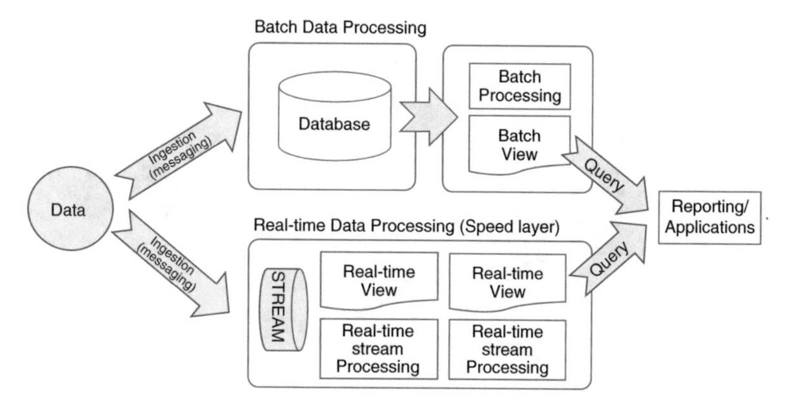

Fig. 6.6 Lambda architecture

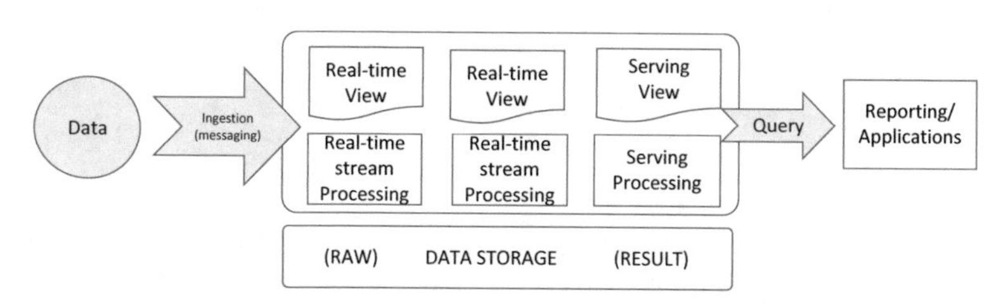

Fig. 6.7 Kappa architecture

The batch processing includes historical data storage and batch job processing such as a pattern discovery, a statistical language modeling, and computations are based on iterating over the entire historical data set.

The real-time processing (called speedy layer) provides low-latency results. It performs updates using incremental algorithms, thus significantly reducing computation costs, often at the expense of accuracy.

The Kappa architecture unifies batch and real-time processing layer into one through processing layer and replacing it with a streaming [10]. Removal of batch processing is possible due to that a batch is a data set with a start and an end, while a stream has no start or end and is infinite. Since a data at rest is a bounded stream, we can consider a batch processing is a subset of stream processing. See Fig. 6.7.

6.3.2 *Stream Processing Systems*

There are few reported distributed stream processing systems that can process Big Data in real time or near-real time. In this section, we will provide a high-level overview of Apache Storm, Spark and stream processing engine.

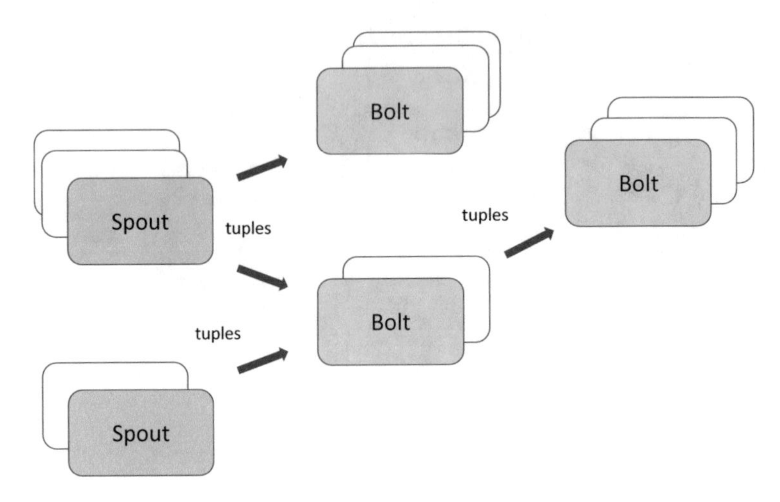

Fig. 6.8 Storm topology

With Storm [11], we can design a processing follow called a topology; a graph of real-time computation implemented with Java programming. A Storm topology is input to the processing cluster. The master node will distribute the code among worker nodes to execute it.

Data is passed between Spouts that emit data streams as immutable sets of key-value pairs called tuples and bolts that transform those streams. Optionally, Bolts themselves can provide data to other bolts.

Spark Streaming [12] is an extension of the Apache Spark API. It slices data in small batches (e.g., called a micro batch) of time intervals before processing them (Fig. 6.8). In the Spark, a stream of data is represented as an abstracted fashion. More specifically, it is an unbounded sequence of tuples; called a discretized stream that is read-only and micro-batch of resilient distributed datasets. It is partitioned collection of records that is processed and created in parallel in a distributed manner.

A spout is considered as a source of data streams in a storm topology. Spouts will read tuples from an external source and push them into the processing topology. When a spout failed to process a tuple, then a reliable spout can replay a tuple. In contrast, an unreliable spout forgets about the tuple as soon as it is pushed to downstream processing.

All processing in topologies is performed in bolts. Bolts are data transformations, and processing functions ranging from filtering, arithmetic operations, aggregations, database joins etc. We can apply multiple steps of bolts to perform complex stream transformation (Fig. 6.9).

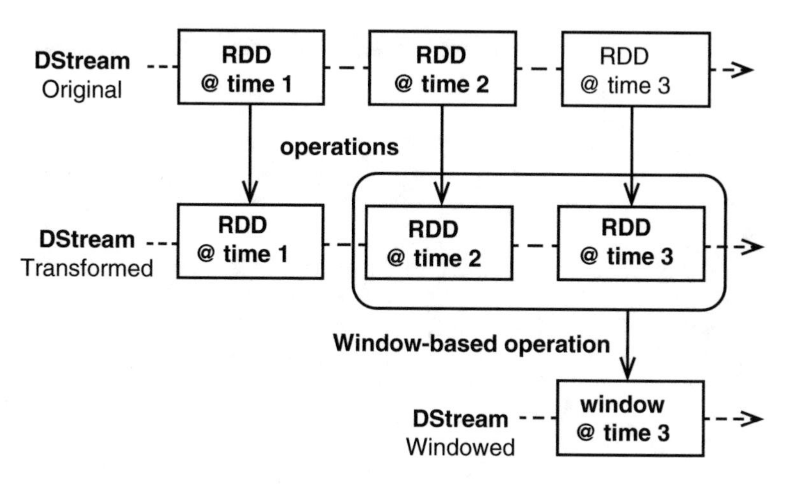

Fig. 6.9 Apache Spark's stream processing

6.3.3 Stream Processing Framework

Time series or streams are lists of data that can be ordered, usually by time. The basic version is a list of triplets:

{<key, timestamp1, type:value1>, <key, timestamp2, type:value2>,…,etc.}

Key is usually the "time series ID," or the "sensor ID"—whatever identifies all the samples/observations as being from the same source. The order comes from the timestamps. Value can be an arbitrary object (e.g., JSON object), however type of value is driving the semantics of operators and what operators can be applied to the object type. For example, a sum of bounded meter data stream totally makes sense. However, a sum of location data stream may not exist.

A modern stream processing system is consist of entity and stream/time series storage service, messaging, and stream processing execution engine (see Fig. 6.10).

Stream storage service: It exposes a simple application programming interfaces for store samples, read samples from the start, to stop, well-studied optimizations to storage possible. Consistent programming interface makes it possible to have a logical data independence. Can also aggressively cache or store older data on disk, hot/newer data in RAM.

Stream processing execution engine: Time series or event databases available today usually focus on being able to record large amounts of data quickly while storing that data in a way that makes it quick to retrieve segments of that data. At query time, they allow users to ask for those segments, potentially with a few aggregation operations to reduce the amount of data return (for example, return a year's worth of data, but only return daily averages). There are a limited number of operations they can perform, usually only the most common aggregation or sampling operations as data reductions. The time series databases do most of these calculations at query time, though they may cache previous results and intermediate data, and might index for the operations they explicitly support.

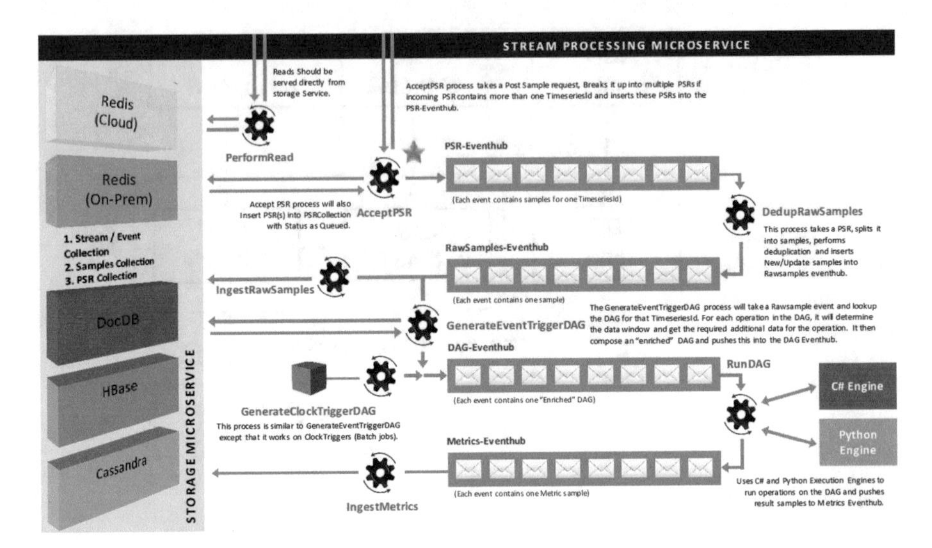

Fig. 6.10 Stream processing system

At its core, most of what industrial Internet of Things applications do with time series data is to transform it into other time series. Aggregate data, "clean" it, enhanced, streams that are fused together, or creating multiple streams from a single input. We can look for problems from rules encoded in logical statements, run statistical tests over the series, use machine learning models to classify segments or to predict future expected values. We may also want to modify machine learning models in light of new data arriving in time series.

Most of what we want to do are run a series of functions over data, potentially as new data arrives, and have the results of those functions available. We do not want to change the original data. In the database community, this is known as a database "view": a function is declared/registered in the database system on top of an existing set of database tables, and then the "view" is queried as though it is also a table, and data is read from the underlying table and returned through the view, potentially transformed.

When data is modified in the source tables for the view (either added, deleted, or updated), with a materialized view, the database system must decide what to do to ensure that any new query to the view is correct—should it update the view on disk, or should it simply note that the view might be incorrect and incorporate the new data from the underlying tables when the view is accessed? If the database decides to update the view on disk, should it delete the entire view and re-compute the whole thing, or should it try to minimize the work done by only changing in the view the data that has changed in the source system? Should the update happen now, or is it better to wait in case there are other updates or if the database is busy on other tasks now? This set of problems is known as the "view update problem."

One way to implement stream processing engine can be thought of as a view engine for stream data and is a solution to the view update problem. A view engine is composed of a declarative view language, designed for stream rather than SQL, and a realization of an engine to instantiate, update, and store views as data arrives.

Declarative views—in JSON, in a DAG/parse tree format representing the graph of the data flow rather than a formal language. Simple enough to edit but yet structured enough to symbolically manipulate, ala a graphical drag & drop flow editor. It defines the inputs/sources, the computation, and the outputs. The declarative view might reference other declarative views as inputs. The analytic job execution layer stitches the graphs together into larger graphs. couldIt supports user-defined functions and user-defined aggregates. Declarative views are DAGs—whole workflow known ahead of time (e.g., Have to define a view before you can query on it).

The DAG structure gives us some optimization possibilities:

- For example, we can combine DAGs if we see two steps, or if we see multiple samples all of which will need similar inputs, or
- If we know the algebraic interfaces of the operators, we can combine nodes in the DAG (e.g., two sums can be done as one step, etc.)
- We can schedule operators to nodes where we know data might already be resident in memory
- Can run aggregations for dashboards, cleaning operations, logical operations for rules and fault detection, machine learning predictions or classifications, call out to external services.
- A stream processing framework should not be to sensor data. It can store event data, image, text, JSON or as a billing engine for consumption or tariff-based billing system.

It is possible that a stream processing could use PostgrSQL, HBase, Document DB, and other NoSQL and back-end data storage. When a new data samples come in, workflow execution engine looks up the DAG and load whatever data is needed, then pass off the fully loaded "Enhanced DAG" to execution engine, then store results. In some cases, we might inject a "clock trigger" to invoke a DAG that needs to be periodically generated, e.g. some alarm calculation [13, 14].

Loading data for the DAG usually only needs a subset of a given series, typically from within the most recent time window (say the past 72 h)—so the lookup of data might consult a cache first before going to the event stream storage system to pull any input data. This unified data access abstraction eliminates separate code base to process data in motion and data at rest.

They also have a de-duplication filter early in the process. If a sample is injected into the pipeline multiple times, they will attempt to remove it to prevent any scheduling of new work.

We could use polyglot style implementation of different execution engines for operators—Python for some, C# for others, could add more—C++ or MATLAB, perhaps. Leverage language features—Pandas in Python has excellent time series statistics features.

The execution engine is responsible for managing security. Views can only reference data that they're authorized to see. We record ownership of views when they're created, and when samples arrive, we ensure that the sender of the sample is authorized to write to that data stream.

6.4 Rule-Based Complex Event Processing

In the previous sections, we discussed stream processing architecture and example systems. One of the common use cases around streaming data is complex event processing that takes multiple event streams as input and makes a decision by referencing registered rules or other analytic tasks. Some event processing rules are as follows:

- When the average electricity consumption increase above 600 KW over any 60 min period, then turn the air conditioner off.
- When two transactions happen on an account from radically different geographic locations within 10 min, then report as potential fraud.
- Forecast electricity demand based on last 6 h and weather forecasts.

CEP is about applying processing rules to streaming event data. An event is a tuple of a time stamp and a value. Examples of this kind of data might be click streams from an e-commerce application, receipts from a transaction, or sensor data. These types of data can be viewed as a constant stream of events. Applying rules on this streaming data enable useful actions and decisions to be taken in response [Better Complex Event Processing at Scale Using a ... (n.d.). Retrieved from https://www.mapr.com/blog/better-complex-event-processing-scale-using-microservi].

Rule-based event processing is well-suited for expressing processing and decision logic in a human understandable language or graphical notation. Thus, it is easy for domain experts to capture the underlying logic operation. Rules should not be compiled but defined declaratively, they can then be modified dynamically at runtime. A Business logic or a decision process is abstracted into rules that are triggered by data and potentially produce new data. Rule management tools usually provide version controls of rules to track the changes and allow a rollback to older versions.

The core of rule-based CEP is, of course, a rule engine. RETE (http://reports-archive.adm.cs.cmu.edu/anon/scan/CMU-CS-79-forgy.pdf) is an algorithm that constructs a matching network called Rete network from the conditions of a set of rules. The matching algorithm in Rete is a method for comparing a set of patterns to a set of values to determine all the possible matches. Rete is a directed acyclic graph that represents higher-level rule sets. They are represented at run-time using a network of in-memory objects.

The Rete algorithm organizes the rules into a tree-like discrimination network sorted according to the left-hand side (LHS) conditional characteristics to be met for the right-hand sides (RHS) of the rules to be executed (see Fig. 6.11).

Alpha node and alpha network:

All the individual patterns to be evaluated for matching are placed into one input alpha nodes. All alpha nodes are connected to the single root node by preliminary type nodes. They separate the patterns to be evaluated based on the appropriate type of facts (data stream) that are entering the system. Two-input beta nodes combine two outputs from one input nodes and construct partial matching results.

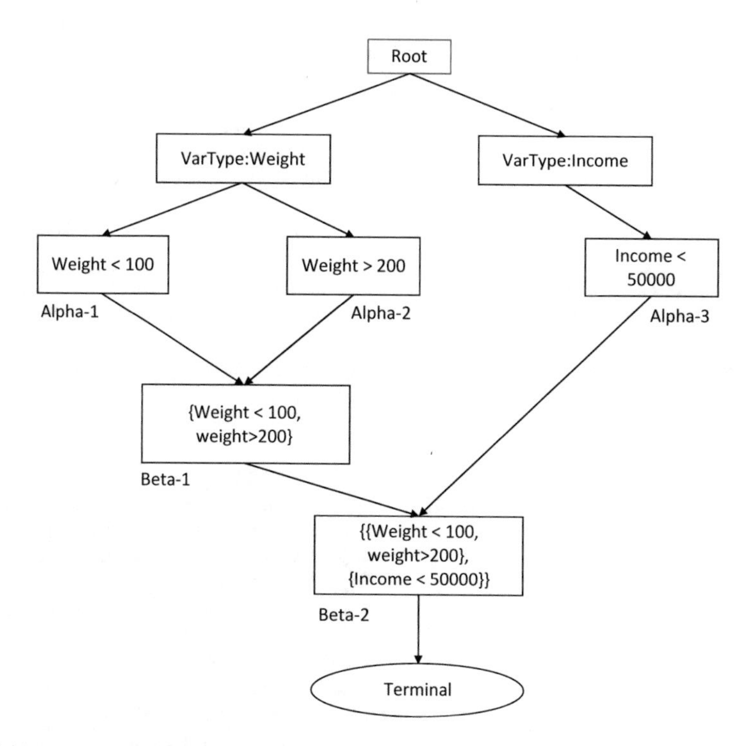

Fig. 6.11 Example of rule representation with RETE algorithm

This network of one and two input nodes assembles at the terminal nodes. Each terminal node represents a rule whose left-hand-side has been satisfied (e.g., all conditions from the root node and to the current terminal nodeRETE are met). By performing this check, the RETE algorithm avoids the multiple evaluations of the same conditions in one decision tree. State memory also prevents the same tests from being repeated on facts that have not been altered.

As Fig. 6.11 illustrates using a simple example rule that considers the weight and income of an individual to make a certain action. All facts (e.g., stream data) enter the network from the root node. The facts are separated according to type by the type nodes, one of which deals with the weight of the user and the other concerns the individual's income. One-input nodes, or alpha nodes, make each test for the conditions to be met off the rules' LHS. Patterns that have been evaluated individually are grouped and joined by beta nodes. Therefore, all three conditions are ensured to be met at once.

Beta 1 node combines the previous one-input nodes testing that the individual's weight is less than 100 or greater than 200. Given the partial matching of Beta 1, Alpha 1 and Alpha 2 can be combined and stored before all three conditions are true. As the final step, the tokens will be verified to pass both the weight and income conditions. Then, they are grouped at Beta 2. It will pass to the terminal node that will initiate related rule checking at the RHS.

Table 6.1 Design consideration in developing complex event processing

Components	Design question
Data ingestion	• Existence of out-of-box data stream data adaptor (e.g., adaptors for Kafka, EventHub, etc.) • Requirement of scale out ingestion, distributed way • Can define ingestion path routing from programmatically or graphically
Define and design rules on the ingested data It enables to decorate custom objects and its properties with validation rules	• Ease to introduce custom decision function as user defined functions (for both bounded or unbounded data stream) • Supporting declarative and imperative validation rules definition • Declarative rule definition with a drag and drop GUI • Supporting composition of validation rules that enables to validate custom object with nested structures • Ease to create your custom validators • Supporting asynchronous validation rules • Supporting shared validation rules • Supporting assigning validation rules to collection-based structures—arrays and lists • Can separate the process of maintaining the application itself completely independent from the process of editing rules that create value for the business
Executing rules	• Supporting agile deployment of rule execution distributed way • Availability of performance monitoring of rule executions • Microservice driven • Ease to deploy
Taking actions per decision when the conditions are met	• Not only making a decision based on facts and rules but also ability to integrate into business process orchestration and integration, therefore, action can be triggered to a target system or people

When it comes down to the implementation phase, we must prepare answers for the following design questions (Table 6.1):

6.4.1 Common Techniques for Identifying Meaningful Events

Often an initial set of rules (e.g., called feature engineering) are provided by domain experts and business analysts. As we acquire more data, now it is time to incorporate machine learning/data mining into stream processing eco-system. Figure 6.12 illustrates such use cases.

A rule execution engine such as Drools (http://www.drools.org) or openrule (www.openrules.com) takes rules defined in a rule database that usually stores them as RETE network and applies to the incoming data stream for a given bounding

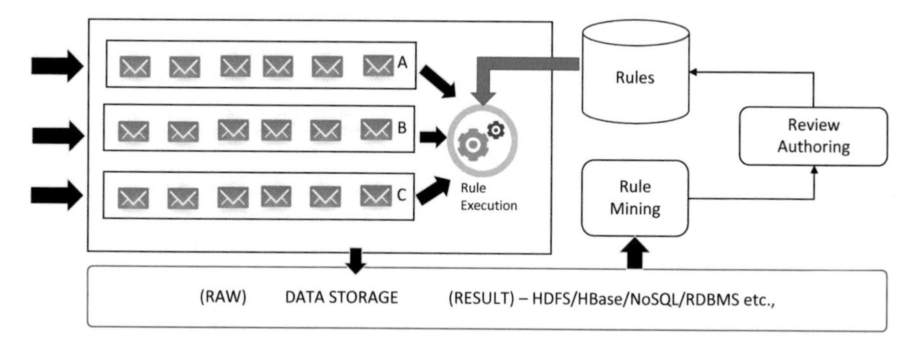

Fig. 6.12 Integration with advanced data mining techniques such as rule mining algorithm (e.g., Associating Rule Mining)

interval. Stream processing's stream writer pushes stream into raw storage. Rule mining process performs relevant events and rule discovery based on scheduled execution plan. Results of discovered rules are presented to review the application for the final deployment to rule database.

Association rule mining is the most common approach to search association among heterogeneous data types (numerical and non-numerical data) [15]. It uses Apriori algorithm is to find items that are frequently associated together [16].

Consider building maintenance optimization example that a building service company needs to service thermostats, valves, actuators, variable air volume (VAV) devices, air handling unites and chillers. In general, chillers supply cold air that is conditioned and distributed by air handling unit. Variable air volume devices are connected to a thermostat that will control the amount of air to be delivered from a VAV. The owner of the company might want to discover which of these components are related to or likely to serviced together. Imagine, we create a declarative view of our stream processing pipeline that may hold 30 days of service log data (Table 6.2).

There are three metrics derive associations among items. The first is "support." It measures how a popular an item set is, as measured by the proportion of transactions in which an itemset appears [17]. For example, *support* (VAV) is 6/9 and *support* (VAV, Thermostat) is 4/9.

The "confidence" measures how likely item Y is serviced when item X is serviced, represented as (X → Y). Therefore, the confidence is the proportion of observations with item X, in which item Y also. In our example,

$$confidence(VAV \rightarrow Thermostat) = \frac{support(VAV, Thermostat)}{support(VAV)}$$

One disadvantage of the confidence measure is that it does not always represent the importance of an association. Since, it only accounts for how popular item X are, but not Y. If item Y are also very popular in general, there will be a higher chance that a transaction containing item X will also contain item Y, thus inflating

Table 6.2 Examples of service transactions

Service tickets	Service items
Service ticket 1	VAV, Valve, AHU
Service ticket 2	Valve, Chiller
Service ticket 3	Valve, Thermostat
Service ticket 4	VAV, Valve, Chiller
Service ticket 5	VAV, Thermostat
Service ticket 6	Valve, Thermostat
Service ticket 7	VAV, Thermostat
Service ticket 8	VAV, Valve, Thermostat, AHU
Service ticket 9	VAV, Valve, Thermostat

the confidence measure. To account for the base popularity of both constituent items, we use a third measure called the lift.

Lift(X→Y) measures how likely item Y is serviced when item X is serviced while controlling for how popular item Y is. For instance,

$$lift\left(VAV \rightarrow Thermostat\right) = \frac{support\left(VAV, Thermostat\right)}{support\left(VAV\right) \times support\left(Thermostat\right)}$$

The first step to analyze association rules with Apriori is to find support count. We could apply thresholding or filtering for faster analysis (e.g., any item that has a support count less than a threshold, then the item should be removed from the future consideration). The result is presented in the following table. Each itemset identified in this step is the first candidate itemsets. Here, we can apply a minimal support count to remove less relevant itemsets.

Table 6.2 L-1 matrix of the frequent itemset.

Itemsets	Support count
VAV	6
Valve	7
AHU	2
Chiller	2
Thermostat	6

The next step is to generate candidate 2-itemsets that are shown below.

[<VAV, Valve>, <VAV,AHU>, <VAV, Chiller>, <VAV, Thermostat>, <Valve, AHU>, <Valve, Chiller>, <Valve, Thermostat>, <AHU, Chiller>, <AHU, Thermostat>, <Chiller, Thermostat>]. Given, 2-itemsets, we can generate the next matrix called candidate itemset C-2. The following table can be generated by join L-1 and L-1, then transactions are scanned and accumulated (shown as support counts) (Table 6.3).

From the above candidate matrix; C-2, we could eliminate <VAV, Chiller>, <AHU, Chiller>, <AHU, Thermostat>, and <Chiller, Thermostat> after applying thresholding (Table 6.4).

The next step is to compute next candidate frequent item patterns shown in the following table (Table 6.5). This can be created L-2 join L-2 procedure presented in (Agrawal, R. Srikant, R. Fast Algorithms for Mining Association Rules, Proc. of the 20th Int'l Conference on Very Large Databases, Santiago, Chile, 1994).

Now, the next step is to find support count for each itemset per transaction scanning. The final L-3 frequent itemsets are shown in Table 6.6.

We can get L-4 frequent itemset, but it will be pruned since it is not frequent.

The final step is to determine relevant rules. Given L-1, L-2 and L-3, all the candidate itemsets generated with a support count greater than a threshold. Now we can use them to develop strong association rules.

For example, an association rule Rule-1; (VAV, AHU) → Valve: when we service VAV and AHU, it is likely we need to service Valve too.

confidence = support(VAV, Valve, AHU)/support(VAV, AHU) = 2/2 = 100%, therefore we can select this Rule-1.

Similarly, we can accept an association rule, Rule-2 (AHU, Valve) → VAV: When we service AHU and Valve, it is likely we need to service VAV too.

Table 6.3 C-2 matrix of 2 frequent itemsets

2-Itemsets	Support counts
<VAV, Valve>	4
<VAV, AHU>	2
<VAV, Chiller>	1
<VAV, Thermostat>	4
<Valve, AHU	2
<Valve, Chiller>	2
<Valve, Thermostat>	4
<AHU, Chiller>	0
<AHU, Thermostat>	1
<Chiller, Thermostat>	0

Table 6.4 L-2 matrix, itemsets and their support count after thresholding

2-Itemsets	Support counts
<VAV, Valve>	4
<VAV, AHU>	2
<VAV, Thermostat>	4
<Valve, AHU	2
<Valve, Chiller>	2
<Valve, Thermostat>	4

Table 6.5 3-Itemset candidates; C-3

3-Itemsets
<VAV, Valve, AHU>
<VAV, Valve, Thermostat>
<VAV, Valve Chiller>
< Valve, AHU, Chiller>
<Valve, AHU, Thermostat >
<Valve, Thermostat, AHU>

Table 6.6 L-3 matrix; 3-itemsets and their support counts

3-Itemsets	Support counts
<VAV, Valve, AHU>	2
<VAV, Valve, Thermostat>	2

6.5 Summary

In this chapter, we discussed IoT data processing pipeline, more specifically, stream processing. Two software design patterns namely Lambda and Kappa were presented, and their pros and cons were discussed. Stream processing system comprises of (1) time series or events storage and retrieval service, (2) messaging subsystem, and (3) execution management.

Data challenges in streaming processing development and design criteria have been introduced with real industrial examples with a potential solution.

In-depth discussion around rule-based complex event processing has been presented with two core algorithms including RETE and Apriori that are the core of most of the event processing engines out in the market.

Complex event processing is a component of whole stream processing ecosystem that is still evolving.

References

1. (Luckham, 2012) David C. Luckham. Event Processing for Business: Organizing the Real-Time Enterprise. Hoboken, New Jersey: John Wiley & Sons, Inc., p. 3. ISBN 978-0-470-53485-4.
2. (Luckham, 2002) David C. Luckham, The Power of Events: An Introduction to Complex Event Processing in Distributed Enterprise Systems, Addison-Wesley Professional; 1 edition (May 18, 2002), ISBN-13: 978-0201727890.
3. (Luckham, 2003) David C. Luckham and Brian Frasca, "Complex Event Processing in Distributed Systems," Stanford University Technical Report CSL-TR-98-754 (http://pavg.stanford.edu/cep/fabline.ps).
4. (Kafka, 2016) Retrieved from https://kafka.apache.org/
5. (EventHub, 2016) Retrieved from https://azure.microsoft.com/en-us/services/event-hubs/
6. (Babcock, 2002) Brian Babcock et al., Models and issues in data stream systems PODS '02 Proceedings of the twenty-first ACM SIGMOD-SIGACT-SIGART symposium on Principles of database systems, Pages 1–16.
7. (Kafka, 2017) Retrieved from https://www.tutorialspoint.com/apache_kafka/apache_kafka_introduction.htm
8. (Kinesis, 2016) Amazon kinesis (https://aws.amazon.com/kinesis/analytics/
9. (Beam, 2016) Apache Beam, https://beam.apache.org/documentation/
10. (Seyvet and Ignacio, 2016) Nicolas Seyvet, Ignacio Mulas Viela, Applying the Kappa architecture in the telco industry, O'Reilly, May 19, 2016.
11. (Strom, 2016) Retrieved from http://storm.apache.org/
12. (Spark, 2016) http://spark.apache.org/
13. (Vera, 1999) James Vera, Louis Perrochon, David C. Luckham, "Event-Based Execution Architectures for Dynamic Software Systems," Proceedings of the First Working IFIP Conf. on Software Architecture (http://pavg.stanford.edu/cep/99wicsa1.ps.gz).
14. (Michelson, 2006) Brenda M. Michelson, "Event-Driven Architecture Overview," Patricia Seybold Group (http://www.psgroup.com/detail.aspx?id=681).
15. (Martin, 2007) J.P. Martin-Flatin, G. Jakobson and L. Lewis, "Event Correlation in Integrated Management: Lessons Learned and Outlook", Journal of Network and Systems Management, Vol. 17, No. 4, December 2007.
16. (Srikant, 1994) Agrawal, R. Srikant, R. Fast Algorithms for Mining Association Rules, Proc. of the 20th Int'l Conference on Very Large Databases, Santiago, Chile, 1994.
17. (Hill, 2008) Friedman-Hill, Ernest J. "The Rete Algorithm." Jess. 5 Nov. 2008. Web. 23 Mar. 2016.

Chapter 7
Constructing Data Service Platform

The data service platform is the backbone of any IoT ecosystem. This is where all the data converges, and we start making sense out of it. Building a data service platform is not about just standing up a BigData processing instance (e.g., Hadoop ecosystem) or migrating all your data into a cloud environment; it is more about building the comprehensive platform for all your data needs and actions. This also includes security, access control, analytics, data lifecycle management and contextualization. Data service platform sometimes contains appropriate traditional technologies as well. In the last three chapters, we have discussed how device information is registered and delivered to cloud service. In this chapter, we will describe tools and technologies for storage, indexing, processing, and retrieval of connected data including telemetry data, metadata about connected devices and others that may generate during a product lifecycle. We will discuss the following topics in this chapter:

- IoT Data management challenges
- Differences in data platform between traditional Enterprise Application Systems and IoT ecosystems
  - Enterprise Data Warehouse
  - Data Lake & Data Virtualization
- Introducing IoT data management and processing technologies
  - Acquisition
  - Storage and Indexing
  - Real-time vs. batch processing
- Data Service
  - Importance of master data service
  - Modeling and managing metadata
  - Managing time series and common operations
  - Security

S.R. Sinha, Y. Park, *Building an Effective IoT Ecosystem for Your Business*,
DOI 10.1007/978-3-319-57391-5_7

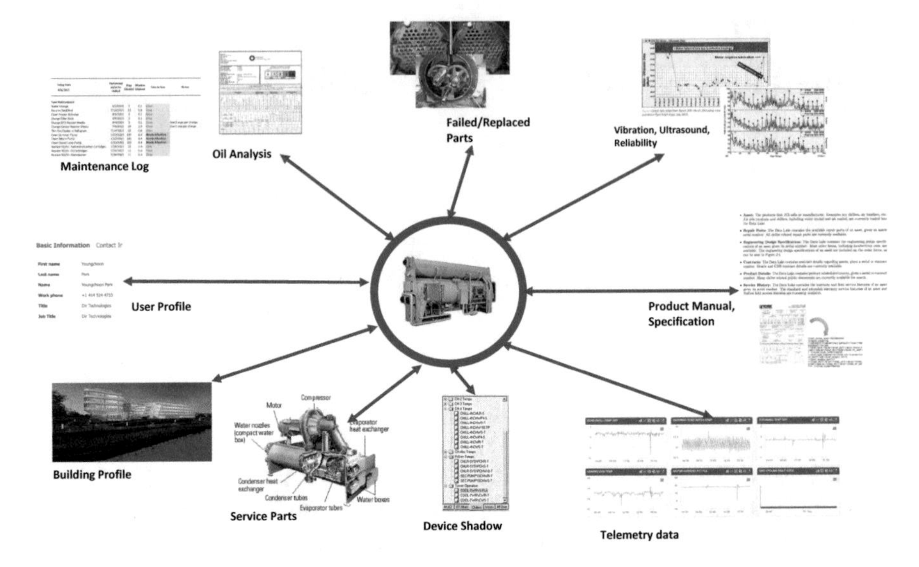

Fig. 7.1 An example of various data types involved in an IoT data management [1]

- Monitoring and managing data platform

Data produced and generated by connected devices and people are essentially big multimedia data by nature consisting primarily of telemetry data, meta-data, acoustic signals (e.g., ultrasound), image, video and audio material as well as text and algebraic notations. Also, text and audio annotations may be incorporated to facilitate content tagging. Analysis, classification, and indexing of IoT data depend significantly on our ability to recognize the relevant information in each of the data streams and fuse it so that the common semantics of all parts will be consistent with the perception of the real world. One of the most challenging problems here is the fusion of the extracted information, among different media and data types. Hence a multi-modal data management with flexible data processing approaches is highly desired to maximize information sharing and to make better actionable decisions with fused information.

There are similar data challenges in developing IoT-enabled solutions as we have witnessed in big data management, namely volume, variety, veracity, and velocity.

Consider a field service application that manages the lifecycle of connected devices and customers. Every service event will generate relevant data for future operational optimizations. Maintenance service events include various data types including telemetry data; high-speed sensor data, a PDF report from oil analysis, raw vibration data, a picture of failed parts, 3D model of product and repair parts, technician's service notes and ultrasound data. A picture of a degraded part can be uploaded to a cloud service for condition assessment. We can use an advanced image analysis service to determine the condition of a part. If a replacement is required, a cloud service will place a replacement part order and a work order. Figure 7.1 illustrates minimally identified multimedia data types for a connected

Table 7.1 Various data types and their usage examples in IoT application

Data type	Usage example
Image	Picture of faulty parts, asset image, condition audit
PDF/scanned document	Product specification, manual, service history
Unstructured text	Service note, customer's problem description
Structured/semi-structured	Application metadata, user profile, business transaction data, etc.
Time series, events	Vibration analysis, faults, sensor reading, safety alert, etc.
Video	Repair sequence instruction, operating instruction etc.
3D CAD drawing	Product design, sub system assembly, replacement part etc.
Audio	Mechanical rotating device operating sample, operating environment noise, etc.

field service application as well as common use cases. Table 7.1 summarizes their examples of usage in IoT applications.

Metadata related to data including capture location, author, time of capture, target asset, etc., must be collected and stored, too. These metadata are critical to contextualize content analysis or telemetry data processing.

IoT solution shall be able to store, index and query various data models including documents, graphs, and events. A predictive diagnostic service will access entity store to find all relevant measurement identifiers to a target asset, then browse time series data and events to create a data frame for analysis, where actual telemetry data is stored in a time series store, which may be a different storage technology.

Predictive failure analytics (e.g., matched potential failures and to develop service recommendations) will examine the data frame to answer "will this asset fail in a week." It will generate tagged events and update asset condition attribute to "high risk or will fail in a week" based on the analysis. Persisting analytic outcomes into a separate time series stream and adding or updating a tag in an entity are common operations from analytic services that will make future causality analysis simple and ease. These entire data analytics and tagging processes usually require multiple storage access to reflect a consistent state, but without ACID transactions from a storage service layer, this requirement can be difficult.

7.1 Differences in Data Platform Between Traditional Enterprise Application Systems and IoT Ecosystems

From data processing perspective, an IoT data service platform shall provide data acquisition, storage, indexing, data processing, master data service, security, etc. IoT data management means more than a real-time sensor data processing. Enterprise-class data service for IoT-enabled business requires a lifecycle management of connected products, associated services, and business partner relationships. An information service platform must be able to capture, correlate and cross-reference all data and events generated by products, services, and interactions

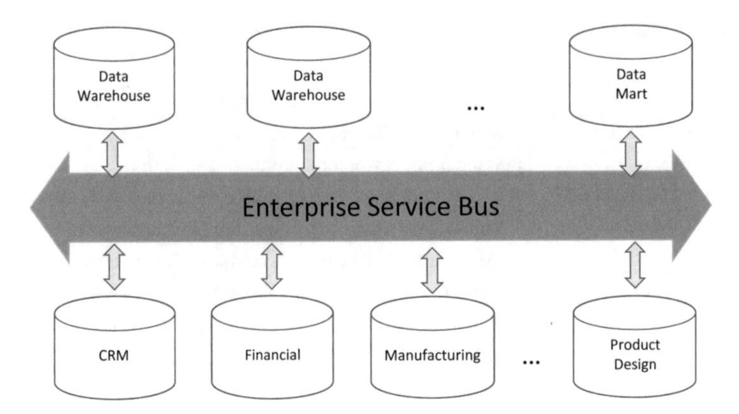

Fig. 7.2 Enterprise application architecture

between a user and a product. We could use existing enterprise applications and data warehouse to perform the IoT data processing. In the next section, we will review enterprise application architecture and describe few limitations of it when it comes to developing IoT solutions.

7.1.1 Enterprise Data Warehouse

It is common to find that enterprise applications such as manufacturing information system, field service management, customer relationship management and product lifecycle management not fully integrated and nor share core information among them. For example, to build a market promotion program, we may need to correlate or cross reference product sales and customers' location data. With the enterprise system architecture shown in Fig. 7.2, we must build a data warehouse solution for this.

Traditional enterprise applications are heavily relying on a relational data management technology that is a highly-designed system. It requires careful design of data repository before the data is stored. This is known as schema on write to support many different types of actives in which canonical data model is required. As we all know, creating a single shared view of the data model used by hundreds to thousands of people and applications is time-consuming and requires a thoughtful data modeling. The following section, we will introduce an alternative technology called a data lake [2].

7.1.2 Data Lake

Often, IoT application requires management and processing of high speed of large volume of heterogeneous data types and data models consisting time series, 3D design, photos, structured, unstructured and semi-structured (e.g., volume, velocity, veracity, and variety). IoT data management is essentially a big data management

Table 7.2 Comparison of IT data platform vs. IoT data platform

	Tradition enterprise application (e.g., data warehouse)	IoT (BigData e.g., data lake with advanced analytics)
Application focus	Transaction management	Transactions and analytics for lifecycle management
Storage and indexing	RDBMS	Distributed File System, NoSQL, Graph, RDBMS, Document, Blob etc. Hadoop eco-system
Architecture	Centralized client and server	Distributed, Shared nothing
Scope of data	Internal data focus	Internal, external and derived data
Data types	Highly structured,	Structured, unstructured, semi-structured, image, time series
Who generate	Mostly human generated data	Machine and human generated data
Volume	GB~TB ranges	TB–PB ranges
Velocity	Medium	High
Data modeling	Schema on Write	Schema on Read
Access	SQU and BI	SQL-like, programming and other options
Data quality	Cleansed	Can be noisy/cleansed
Benefits	Fast and consistent performance	Executes on tens to thousands of servers with superb scalability at lower cost
	High concurrency	
	Easy to consume data	Parallelization of traditional programming languages
	Rationalization of data from multiple source into single enterprise view	Supports higher level programming frameworks
		Radically changes the economic model for storing high volumes of data
	Clean, safe, secure data ONCE THEY ARE DESIGNED RIGHT	
		Running advanced data analytics including deep learning

by nature. Table 7.2 illustrates a comparison between traditional enterprise data management vs. IoT data service platform.

Authors believe that traditional ERP systems and IoT data service platform will converge into a single cohesive data service. An example of such IoT and ERP integrated data service systems from Johnson Controls (http://mysmartequipment.com/) is shown Fig. 7.3. The data service platform acquires not only IT system generated data, but also connected product generated information. It leverages a Hadoop eco-system to implement data lake that helps to create a converged view of all data sources.

Data Lake is a low-cost fault tolerant distributed data store and processing system, where an enterprise can place various data including internal, external, partner's, competitor, business process, social, and people; store all data.

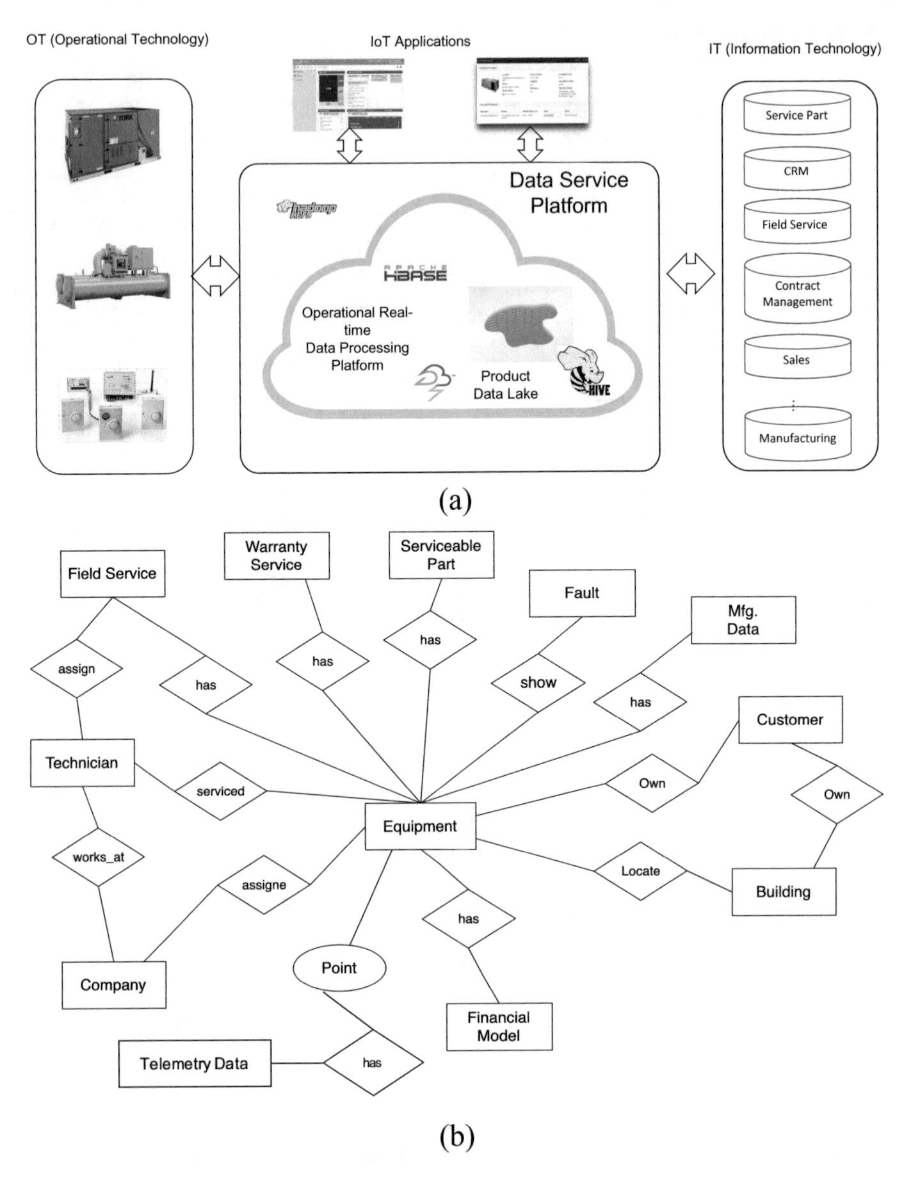

Fig. 7.3 (**a**) An example of enterprise scale IoT application from Johnson Controls [3], (**b**) the connected data model used in connected service application, note that relationships and entities are dynamically updated and/or added

Data Lake provides scalability that can manage a growing amount of data growth linear fashion through distributed data processing platforms. Hadoop is the horizontal scalability tools, which leverages the HDFS storage. One of the main benefits of data lake is a convergence of all enterprise data sources. It can store virtually everything including logs, XML, JSON document, multimedia, sensor data, binary, social data, chat, and people data. Kafka, Storm, and Flume are technology options to ingest the high-speed and large amount of data.

Flexible data processing enables agile and intelligence from the structured and unstructured data, which is stored in the Data Lake we should implant the schema for the data and make the data flow in the analytical system [9 Key Benefits of Data Lake - Data Science Central. (n.d.). Retrieved from http://www.datasciencecentral.com/profiles/blogs/9-key-benefits-of-data-lake]. Data lake eliminates the need for schema design at the time of ingestion; we can do it at the time of data reading (a.k.a. schema-on-read).

The Data Lake leverages the NoSQL, distributed file system, JSON store and other data storage technologies to store the data based on schema-less write and schema based read mode. This schema-on-read is very much useful at the time of data consumptions and handling various data format. Once the data is ingested, cleansed, and stored in a structured SQL storage of the Data Lake such as HIVE, we can reuse the existing SQL scripts to retrieve data and create views.

The Data Lake shines at utilizing the availability of a large amount of data with distributed computing (e.g., move code close to data) with advanced analytics including deep learning algorithms to classify items of interest that will power real-time decision analytics [4].

7.2 Introducing IoT Data Management and Processing Technologies

In the previous sections, we learned various challenges in IoT data management. In this section, a reference architecture is presented in detail. The reference model includes various technology components including data lake, master data management, knowledge extraction and indexing, and various data ingestion techniques. Presented reference architecture suggests various abstraction components prevent technology and/or vendor lock-in situations. Figure 7.4 represent our reference architecture of IoT data service platform. Throughout this chapter, we will discuss features and implementation details for each component.

7.2.1 Storage Layer

Two crucial characteristics of data are the shape (or model) of the data and the semantics of the data.

Depending on the "shape" of the data or the data "model", different data management systems may be better suited for that data than others. For example, event data might be best stored in a key-value store, whereas semi-structured data might fit well in a document-oriented database such as MongoDB. Other data shapes

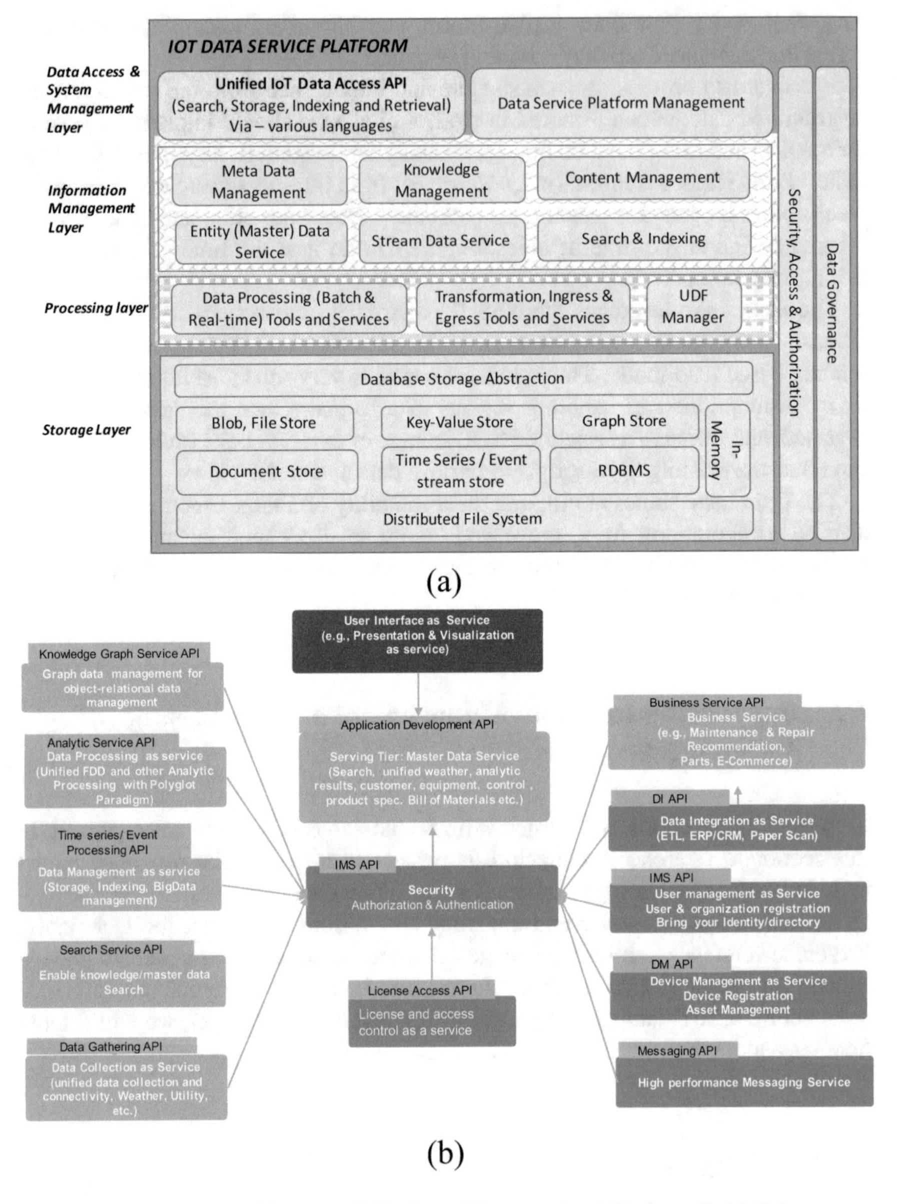

Fig. 7.4 (**a**) Reference Architecture of IoTatform, (**b**) companion APIs for unified IoT data access

might include graph data, relational and tabular data, multimedia data such as audio and video where the raw readings are stored in an object store with associated metadata and index data stored in a second system. The Data Hub provides applications with the appropriate system: a legacy application might expect a traditional filesystem or access to a specific Relational Database. A web application or application running in a PaaS may be able to select from several different relational databases.

Of course, an application may have actually have multiple data types to store, with no one system being ideal for all types. The storage layer embraces the notion of "polyglot persistence" [5], that is, to simplify the storage of heterogeneous data types and various data models into different data stores, allowing for applications or components of applications to choose an appropriate system based upon the way data is being used. Polyglot persistence brings with it new challenges, particularly around locating and using data.

7.2.1.1 Processing Layer

Processing layer consists of data processing platform for both real-time and batch processing and data movement and transformation. Apache Hadoop is a dominant player in this space. Hadoop ecosystem now includes both real-time processing frameworks such as storm and famous MapReduce for batch processing. Time series processing and analytics are also part of this layer. Cluster resource management and application distribution are critical components of data processing platform.

7.2.2 Information Management Layer

Information management layer comprises of various data service including master data service, metadata management, knowledge management (e.g., knowledge graph), search service, etc., each service in this layer uses storage abstraction API to maintain service specific data.

7.2.3 Data Access and System Management Layer

This layer provides unified IoT data access APIs and platform performance management tools. Unified data access APIs provides secure access and CRUD (Create, Read, Update and Delete) operation on entity, meta-data, stream, and other IoT application relevant internal and external data stored in data lake. Data platform monitoring service shall provide at least the following metrics and features.

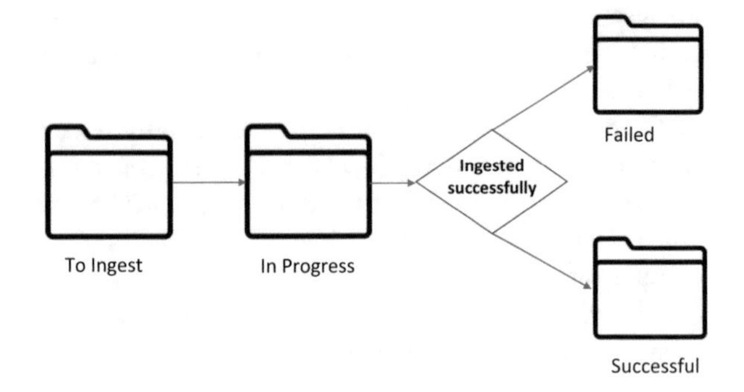

Fig. 7.5 Four different locations for a reliable data ingestion

7.2.4 Data Acquisition: Transformation, Ingress & Egress Tools and Services

In Fig. 7.5, transformation, ingress & egress tools, and service refer data acquisition. In enterprise application setting, most of source systems have been developed independently over many years and based on different data management technologies, there is no unified data acquisition interface to acquire IT systems data into to a platform. This is the common circumstance for most enterprise data integrations. This cause some complexity in data ingestion into our IoT data platform and requires different architectural and toolchain decisions depends on data types and ingestion scenarios. There are few key factors drive ingestion architecture and tools. They are the type of data sources, data transportation protocols, and update patterns.

Data source: in enterprise IoT solution, we must able to bring data from databases, event streams, files, and web services, etc. The data source will derive read speed and pull vs. push the decision.

- Data transportation protocol: streaming data could be delivered through various protocol standards including MMS, MQTT, SMS, CoAP, etc.
- Update pattern: the incoming data could be appending, change set, and replace.
- Timeline requirement: How fast/often data need to be loaded?
- Data Structure: will the data need to be transformed in flight or additional processing?
- Target data storage: what target storage system will be used to store ingested data.

7.2.4.1 File-Based Batch Data Ingestion

Even though there are many options in bringing data into IoT data service or data lake, still, file transfer protocol (FTP) and shared file system, are popular file based batch mode data ingestion. In general, four different staging areas for data ingestion, namely ToIngest, InProgress, Failed, and Successful are prepared. All data files or records to be ingested into the data platform must be placed into the ToIngest area. As files are being ingested into a data platform, files are moved into the inProgress area. Successfully loaded files are moved into the Successful area.

All the tools discussed in this chapter can be classified as either pushing or pulling tools. The important thing to note is the actor in the architecture because, in the end, that actor will have additional requirements to consider, such as:

- Bookkeeping of what has been sent
- Handling failures; including retries or failover
- Minimize system performance impact of the source system that data is being ingested from
- Access and security, non-human account provisioning with read-only access

7.2.4.2 Data Flow-Based Multi-purpose Data Ingestion

In this section, we will discuss more comprehensive and extensible data ingestion and flow processing tools for data movement, processing, and translation between systems.

Flume: It is distributed system for collecting log data from many sources, aggregating it, and writing it to Hadoop Distributed File System (HDFS). By design, it is reliable and highly available while providing a simple, flexible, and intuitive programming model based on streaming data flows. Flume maintains a primary list of ongoing data flows and uses Zookeeper to keep the redundant copies of the list to manage failure scenarios. [6], A Flume agent is a Java-based process that hosts the components through which events flow from an external source to the next destination (hop).

A Flume source consumes events or data delivered to it by an external source like a web server or data readers. The external source sends events to Flume in a format that is recognized by the target Flume source. For example, Thrift [7]. Flume Source to receive events from a Thrift Sink or a Flume Thrift RPC Client or Thrift clients written in any language generated from the Flume thrift protocol. When a Flume source receives an event, it stores it into one or more channels. The channel is a passive store that keeps the event until it's consumed by a Flume sink [6].

The sink reads and removes the event from the channel and puts it into an external repository like HDFS (via Flume HDFS sink) or forwards it to the Flume source of the next Flume agent (next hop) in the flow. The source and sink within the given agent run asynchronously with the events staged in the channel (Fig. 7.6).

Chukwa: Apache Chukwa is built on top of the HDFS and Map/Reduce framework, therefore it inherits Hadoop's scalability and robustness. Apache Chukwa also includes a flexible and powerful toolkit for displaying, monitoring and analyzing results to make the best use of the collected data [8]. Unfortunately, writing MapReduce jobs to process logs is somewhat tedious, and the batch nature of MapReduce makes it difficult to use with real-time data stream. Furthermore, HDFS does not support appending to existing files. It provides a scalable distributed system for monitoring and analysis of log-based data. Some of the durability features include agent-side replying of data to recover from errors.

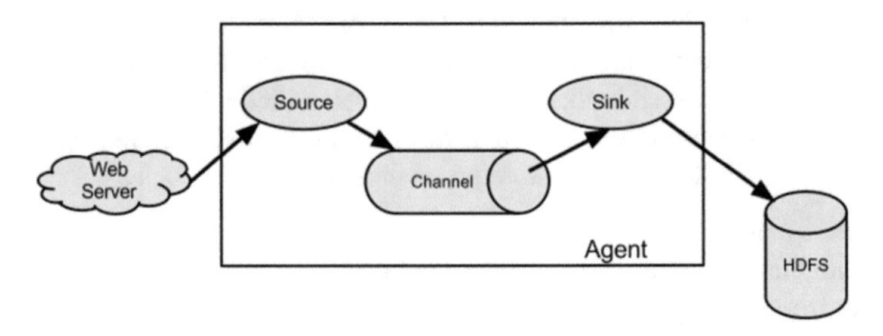

Fig. 7.6 Data flow processing in flume

Gobblin: It is intended to be a unified data ingestion framework developed by the LinkedIn team. Gobblin [9] provides a uniform interface to ingest data from a variety of data sources, e.g., databases, rest APIs, FTP/SFTP servers, files, etc., onto Hadoop. Figure 7.7a illustrates how a unified data ingestion works with Gobblin that handles the common data ingestion ETLs, including job/task scheduling, task partitioning, error handling, state management, data quality checking, data publishing, etc. Ingestion uses the same execution framework regardless of data sources and manages metadata of different sources all in one place. Users can add new adapters for new data extraction and sources in any development settings. A Gobblin job is constructed from a set of work units or tasks (see Fig. 7.7b). A work unit is responsible for extracting a portion of the data. The tasks of a Gobblin job are executed by the Gobblin on the deployment setting of choice (e.g., a single node, Hadoop MR, Yarn in Fig. 7.7b). The Gobblin runtime is responsible for running user-defined Gobblin jobs on the deployment setting of choice.

In summary, a Gobblin job creates and runs data ingestion tasks, each of which is responsible for extracting a portion of data to be pulled from the job. A task composes the Gobblin constructs into a flow to extract, transform, checks data quality on, and finally writes each extracted data record to the specified sink.

NiFi: It supports scalable directed graphs of data routing, transformation, and system mediation logic. It is more big data processing flow modeling and runtime platform. The core concept of NiFi is to provide data flow execution platform to solve data movements, processing, and transformation between systems. It comes with a graphical user interface to author data flow processing and monitoring tools. It has a powerful flow expression language to describe processing flow. FlowFile is a container to abstract all NiFi's data elements [10]. A FlowFile consists of content and attributes. The content section of the "FlowFile" describe the data on which to operate. For example, if a file is picked up from a secure file transfer location using the Get SFTP Processor, the contents of the file will become the contents of the FlowFile.

The attributes section of the FlowFile describes information about the data itself, or metadata. Attributes are key-value pairs that represent metadata and routing and processing the data related information (Fig. 7.8).

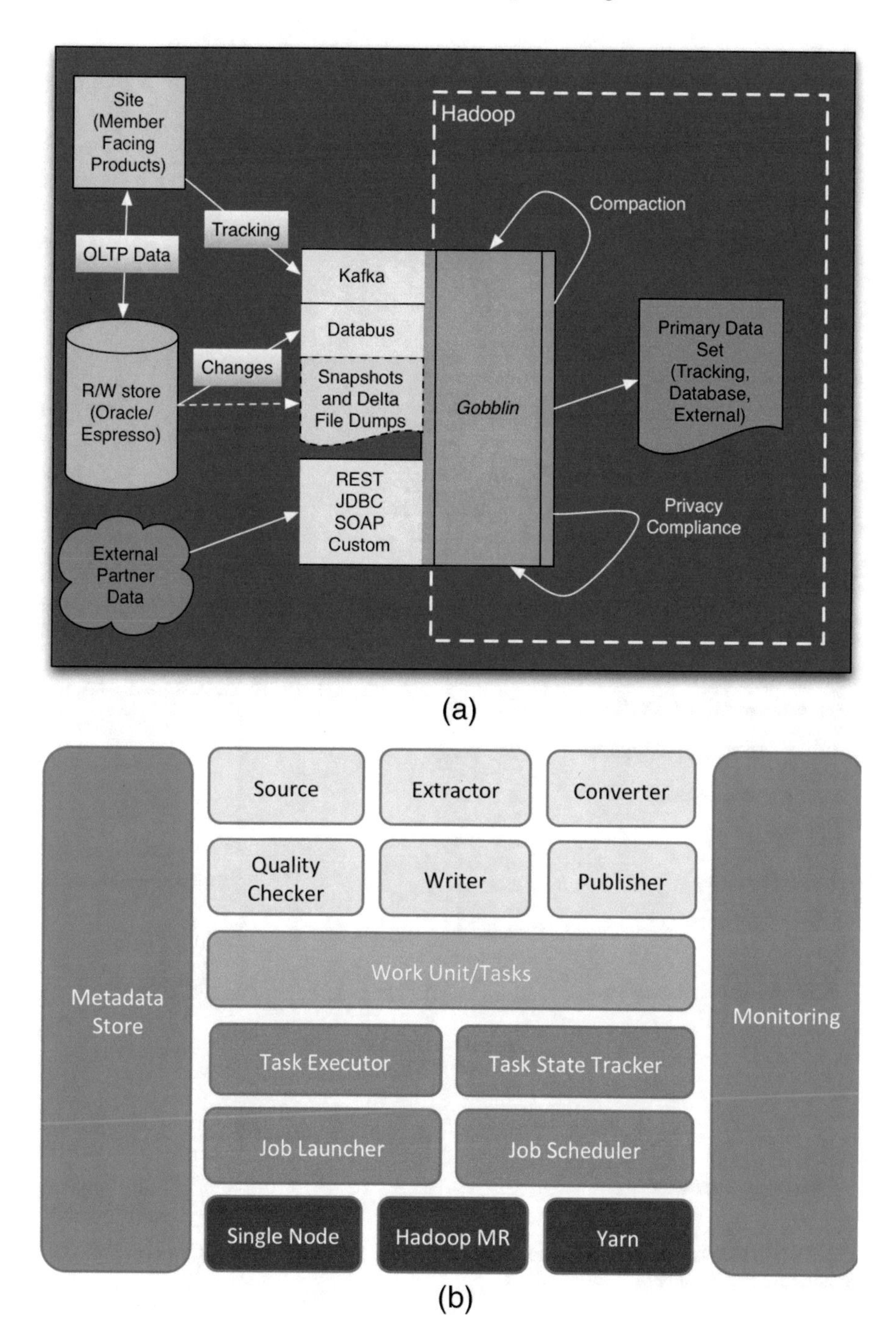

Fig. 7.7 (**a**) High-level operational overview of unified data ingestion with Gobblin. (**b**) Gobblin architecture overview

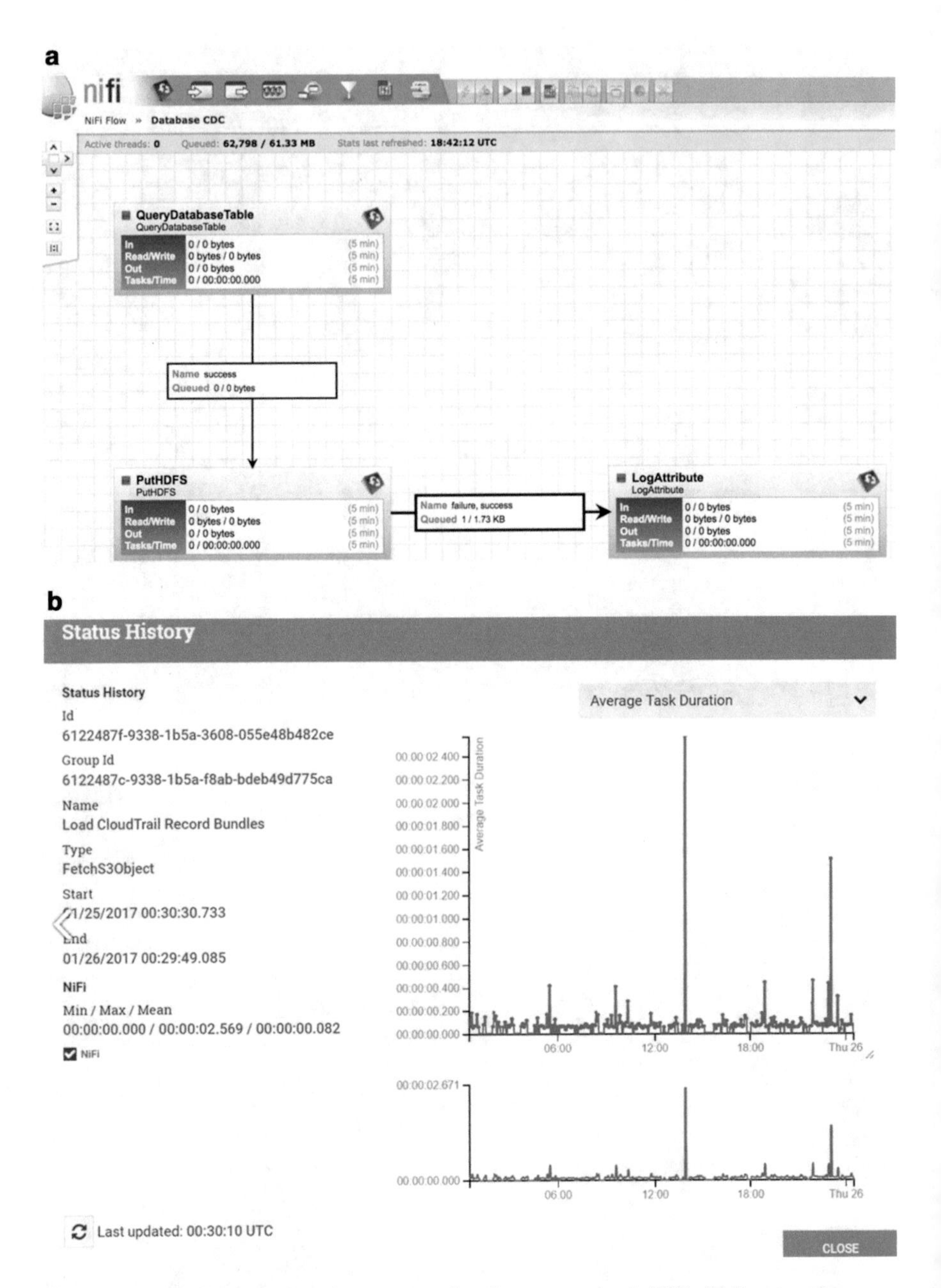

Fig. 7.8 (**a**) Graphical user interface to create data flow processing in NiFi. (**b**) Event performance monitoring tool from NiFi

Table 7.3 Tool selection guide for and data Ingress

Type from	Ingestion type	Ingestion timing	Ingest to multiple locations	Tools
Unstructured file	Bulk load	Non real-time	No	FTP, File Share, Apache NiFi
RDBMS	Bulk load	Non real-time	No	Sqoop, StreamSet
RDBMS	Bulk load	Non real-time	Yes	NiFi, StreamSet
Stream	Small message (semi structured)	Real-time	Yes	Kafka or Messaging
				Queue, StreamSet, Flume, Kafka
RDBMS	Incremental (or change)	Near real-time	Yes	NiFi with additional
				DBMS tool such as GoldenGate, DbVisit
Log file	Bulk and incremental	Near real-time	Yes	Flume, Kafka

Other Data Ingestion Tools

Sqoop

Apache Sqoop [11] is designed for transferring bulk data between Apache Hadoop, more specifically HIVE and HDFS and structured datastores such as relational databases. It allows bi-directional data replications with both snapshots and incremental updates. Sqoop is a pull-based solution that must be provided various parameters about the source and target systems, such as connection information for the source database, one or more tables to extract, and so on. Given these parameters, Sqoop will run a set of jobs to move the requested data (Table 7.3). Detailed usage can be found from https://sqoop.apache.org/docs/1.4.6/SqoopUserGuide.html, and few important sqoop based ingestion patterns are discussed in [12].

7.3 Storage and Indexing

Figure 7.9 illustrates an example of multi-modal information visualization that is a highly efficient IoT data analysis tool to understand causalities of events collected from various sensors, applications including service logs (e.g., technician's notes), vibration analysis, oil analysis, cameras, ultrasound sensors, thermometers, and weather stations [13].

A variety of data types and data models in IoT domain introduces a set challenges in storage and indexing of them. We could employ "polyglot persistence"

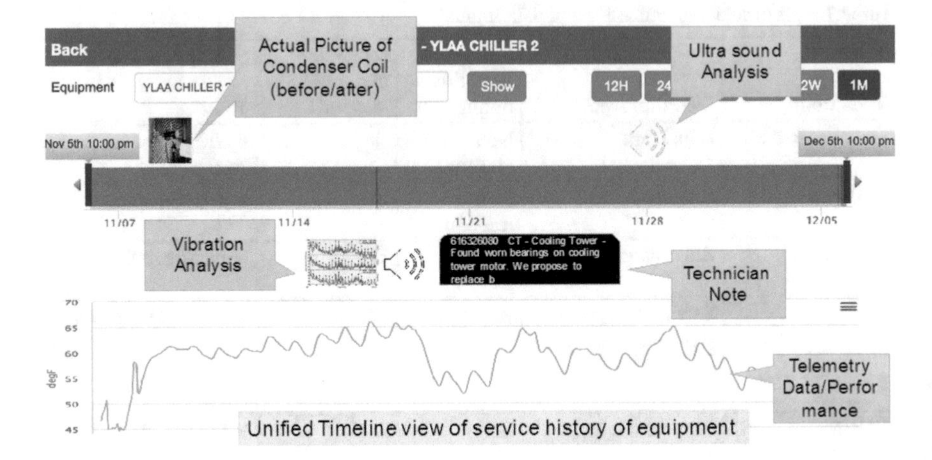

Fig. 7.9 A unified timeline visualization of failure, repair and operation, failure and other related events and telemetry data stream

approach when we want to store heterogeneous data types and various data models, since it suggests to use multiple data storage technologies, chosen based upon the way data is being structured and used by individual applications or components of an application [5].

In this section, we will use a field service application as an example to illustrate data modeling, storage, indexing, and processing challenges in developing an IoT data service platform.

Application data can be modeled with JSON like semi-structured objects or structured entities that can be efficiently stored and queried with relational database systems. Examples of application data are a description of installed location of an asset, owner information details, product specification, firmware version, and telemetry data points, etc. In field service example, maintenance and repair events and service recommendations (e.g., a result of predictive analytics for example) for an asset can be stored in the key-value store.

Completed service histories can also be presented as a document and be stored in a document database (e.g., maintenance transaction summary and details). Static and dynamic relationships among entities including owner information, locations, asset details and other maintenance recommendations (i.e., a result of predictive analytics) can be modeled with Graph stores.

As a system learns and discovers more about relationships among events and entities, the system consistently introduces new relationships, update or delete existing relationships through analytics services (i.e., enriching semantic relationship). For example, a newly added maintenance event shall lower a future failure mode of an asset by updating causal relationship between an asset and a failure type. A set of recommended maintenance services (e.g., a set of entities) can be introduced to an asset by creating or updating a relationship between an asset and a service basket. We will discuss in detail how this dynamic relationship management

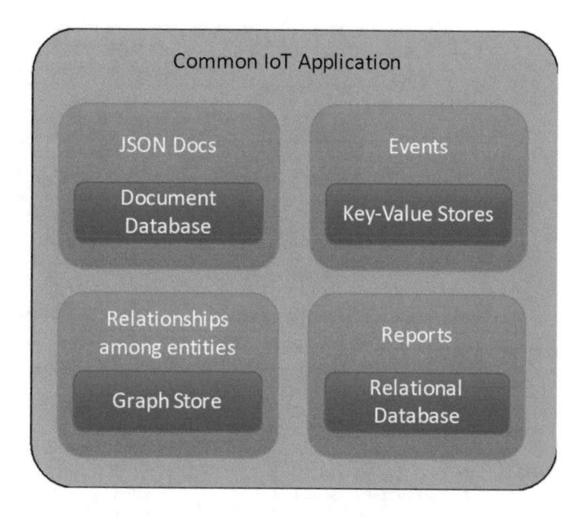

Fig. 7.10 Common IoT applications require multiple data models and storage technologies

could be done. Figure 7.10 illustrates various data models and storage technologies to develop an IoT solution.

With polyglot persistence paradigm, IoT application is responsible for proving ACID (Atomicity, Consistency, Isolation, Durability) of among different data models and storages. We consider polyglot persistence is a workaround for heterogeneous data modeling, indexing and storage problems, not a solution to them.

7.3.1 *Example: Data Models for an IoT-Enabled Field Service Applications*

A field service application will be employed to describe and demonstrate the challenges in data management and processing when we develop an IoT solution. It involves heterogeneous data types, various storage technologies and various data models by nature.

A connected device shall be modeled with a digital twin that is a virtual representation of a physical device, where a digital twin is a computerized companion of a physical asset. It can be 3D CAD model with product specification or a set of telemetry data points associated with an asset. A common data model to represent a digital twin is a document (e.g., JSON-based document), that can be managed with document stores including MongoDB, Cassandra or DocumentDB. A special back-end service performs state consistency between a physical device and a device twin.

Set of applications specific or business data including a location of an asset, a product operating specification, an owner information of assets, an organizational hierarchy of assets, a service provider details and other information to perform predictive field services must be modeled and stored too. In general, entity relationship modeling is popular, where entities can be stored in a relational database or a document database, where semantics between entities must be handled by an application. A graph database is also an option to model dynamic relationships among entities.

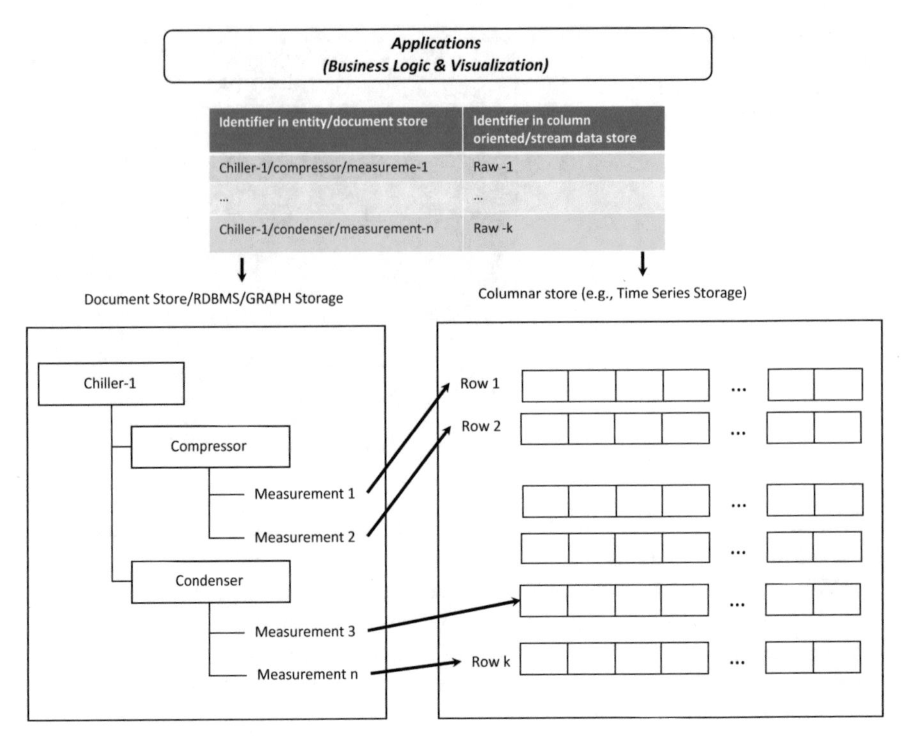

Fig. 7.11 IoT application is responsible for maintaining ACID properties between two different storage technologies

A connected device generates many different types of streaming data including sensor readings, click streams, and others. Therefore, streaming data management and processing are an essential part of any IoT application and platform.

7.4 Unified IoT Data Management and Processing

In previous sections, we have presented challenges in IoT data management and processing. With "polyglot persistence" principal, our IoT application data management must create a mapping to provide strong consistency of data stored in two different data stores namely entity store and telemetry data store (see Fig. 7.11 for details).

Maintaining mappings and developing custom ACID services per applications will be expensive and tedious. A better way to handle this is to build a right set of abstractions that provide APIs for application developers and data management applications.

A proposed reference architecture is shown in Fig. 7.2. Various data storage technologies are abstracted via storage I/O abstraction that provides consistent CRUD (Create, Read, Update and Delete) operations across multiple storage technologies.

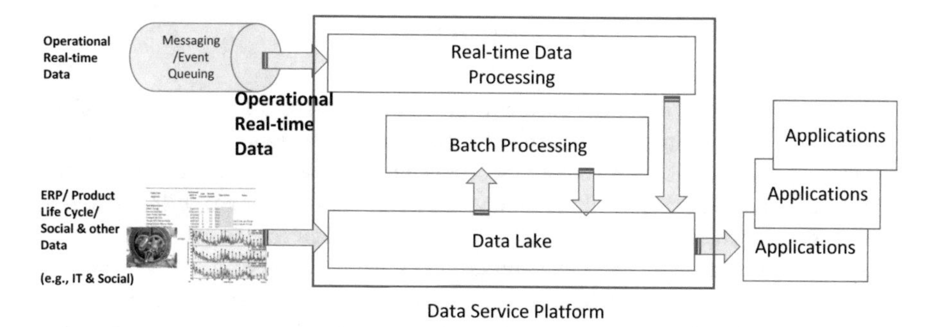

Fig. 7.12 A commonly referenced IoT Data Processing System (a.k.a., lambda architecture) with Hadoop-based data lake. Often it is criticized that it requires two code paths (i.e., real-time and batch) for the same analytics

Entity management provides master data service on stored entities and unified CRUD operations via storage abstraction APIs. Multimedia data and stream management provide similar functionalities of entity management while it deals with various media types, blobs, and files. A set of registered analytic services for time series, image analysis, and other IoT data processing service should be available to perform scheduled tasks.

ACID management shall maintain consistency among entities, attributes of entities, and events and/or telemetry data. Any changes in entity store will trigger consistency check service to make updates on other storages (e.g., the foreign key relationship among different data stores). Through the proposed reference architecture, the application does not need to maintain tedious mapping anymore and to interact with various low-level storage interfaces.

7.4.1 Real-Time vs. Batch Processing

Figure 7.12 illustrates a popular lambda architecture for IoT data processing. It takes incoming sensor data readings shown as operational real-time data, received from a messaging system such as Kafka (see Chapter 6 for details), and route them to storage for batch processing and to stream processing pipeline for real-time processing.

There are many reported IoT solution examples based on the lambda architecture, however, a true enterprise-level IoT solution requires many other data processing and service components to make a complete solution.

In this section, we will review few popular technologies for real-time processing and batch processing on Hadoop-based data lake infrastructure.

7.4.2 Batch Processing Framework

A well-known batch processing framework is MapReduce that is a programming approach. A MapReduce job runs in the background of Hadoop to provide scalability and easy data-processing solutions. MapReduce breaks down a task into small parts and distributes them to other computers. Later, the results are collected at one place and integrated to form the result dataset. The MapReduce job contains two important tasks, namely Map and Reduce. The Map task takes a set of data and converts it into another set of data, where individual elements are broken down into tuples (key-value pairs). The Reduce task takes the output from the Map as an input and combines those data tuples (key-value pairs) into a smaller set of tuples. The reduce task is always performed after the map job. The MapReduce job is processed through multiple phases namely input phase, map phase, intermediate keys, optional combiner, shuffle and sort, reduce phase and output phase.

In Input phase, a record reader translates each record in an input file and sends the parsed data to the mapper, where parsed data is in the form of key-value pairs. Map is a user-defined function, which takes a series of key-value pairs and processes each one of them to generate zero or more key-value pairs. They key-value pairs generated by the mapper are known as intermediate keys. An optional combiner is a type of local reducer that groups similar data from the map phase into identifiable sets. It takes the intermediate keys from the mapper as input and applies a user-defined code to aggregate the values in a small scope of one mapper [14].

In Shuffle and Sort phase, The Reducer downloads the grouped key-value pairs onto the local machine, where the Reducer is running. Key sorts the individual key-value pairs into a larger data list. The data list groups the equivalent keys together so that their values can be iterated easily in the Reducer task. In the Reducer phase, The Reducer obtains the grouped key-value paired data as input and runs a Reducer function on each one of them. Once the Reducer processing is completed, it generates zero or more key-value pairs to the output step. Output Phase translates the final key-value pairs from the Reducer function and writes them to a file using a record writer, where Reducer matter is involved. A detailed example and tutorial of MapReduce can be found from http://www.drdobbs.com/database/hadoop-writing-and-running-your-first-pr/240153197.

7.4.3 Real-Time Processing Frameworks

Real-time analytics refers both interactive real-time data analytics and real-time stream analytics. Real-time interactive analytics means users can issue ad-hoc queries to a large data set. Queries should respond within 10–12 s, which is considered the upper bound for acceptable human interaction. Few example data management technologies are Hana from SAP, VoltDB, MemSQL, and Apache Drill.

As we discussed in the previous chapter, the Apache Hadoop ecosystem has become a preferred platform for IoT real-time stream data processing. Apache

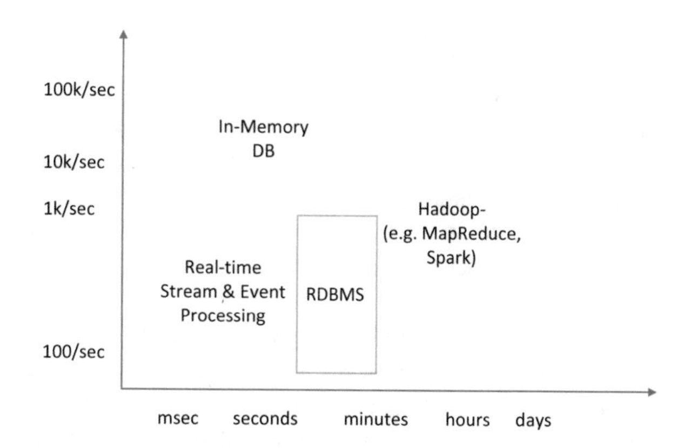

Fig. 7.13 Time to act vs. number of events handled by technologies

Kafka, Apache Flume, Apache Spark, Apache Storm, and Apache Samza are increasingly popular in IoT data transmission and processing. They belong to real-time stream processing category in Fig. 7.13 that describes relative performance of different data processing technologies [15].

Few architectural design patterns for real-time streaming analysis are presented in [16] and functional uses cases are discussed in [15].

7.5 Data Services

Once the shape and data storage system are decided, an equally important set of concerns are the semantics of the data. Applications and systems must be able to agree on the semantics of data in order to interoperate. In some cases, the data may be marked up in well-known standards such as the Brick Schema [17] and Haystack [18] tags for built environment standards or the PADIS XML [19] schemas for airport data. In other cases, the semantics are meaningful by convention: a stream of sensor data readings or a file with the schedule of upcoming facility repairs.

Data Service is a way to store and manage the semantics of the data stored and flowing through the Data processing, Master Data Service and relationship discovery, and to provide a way for applications, users, and system components to discover and use that data. The Data Service is a set of APIs, design patterns, and best practices When data is stored in the Data Service, the semantics of the data are stored as well. For example, a business analyst accessing a dataset can find information about the schema of the dataset beyond the column names and types.

Data may need to be transformed into a different form before being consumed by an application: if a user discovers a data set they wish to use but their application can only consume building data using the Brick schema, if such a transformation is available the Digital Service can provide a link to the translated data.

Metadata in the Data Service is managed as a graph of entities and relations, and new relations and metadata about data can be added to the Data Service Storage over time and consuming applications will automatically be able to take advantage of it.

7.5.1 Importance of Master Data Service

In this book, we refer master data as an entity that is relevant to a domain of interest. An entity is a data model or a set of records that users can create, read, update, and delete. You can introduce relationships among entities, therefore you can relate to other entities in one entity based on semantics. For example, you could create a connected product description as an entity and relate to service events to track all service or failure histories.

Master data service requires integration of people, business process, and tools. Business must have an owner of the data definition and standard schema development and revision processes. The tool must be flexible to maintain multiple version and incremental change management. One implementation could use JSON storage or document database to store master data. We can prepare two columns; one is golden record column (e.g., MASTER_CF in Fig. 7.14) and another is a raw column that holds non-cleansed raw data. Key can be a composite one that contains source system's record identifier. This enables ease of referencing raw data from a golden record.

The master data service allows you to securely store and manage data within a set of standard and custom entities. To create a custom entity, we must register and maintain entity schemas along with versions. An entity is a set of attributes used to store data similarly to a table within a database.

Some of the benefits from having common entities/master data service are

- Easy to manage—The metadata and data are stored and retrieved through unified IoT master data service API. Developers don't need to know about the details of how they're stored (see Fig. 7.2).

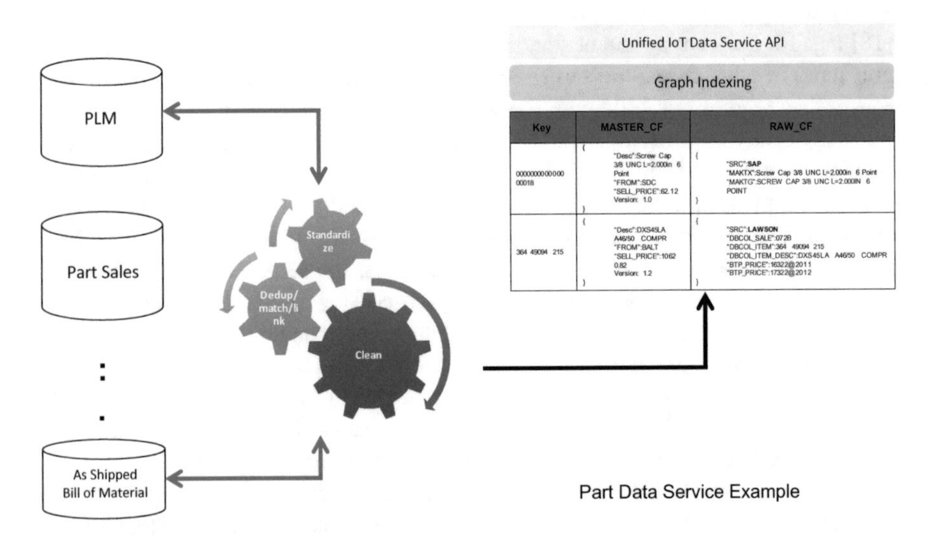

Fig. 7.14 Part master data service example

- Easy to share—You can share semantically clarified data among applications.
- Easy to secure—Data is securely stored via partitioning and role-based access control, so that users can see it only if you grant them access. Role-based security allows you to control access to entities for different users within your organization.
- Simplifying metadata management—Data types and relationships are leveraged directly within IoT application.

For example, suppose we have connected field service data model shown in Fig. 7.15 and trying to build two applications A and B. A could be a fault analysis dashboard and B could be service billing tool. Both applications share entities; technician, equipment, and building. With master data service, the application accesses them via entity access API. Therefore, developers don't need to worry about underlying storage and indexing. Application users will maintain the single truth of shared entities.

7.5.2 *Managing Time Series and Common Operations*

In the JCI's example, a connected device failure can be modeled and predicted from a nightly schedule batch process implemented as MapReduce jobs. Apache Storm is being used in abnormal performance analysis for real-time reporting. A system health report includes not only failure potential but also suggested maintenance and repair

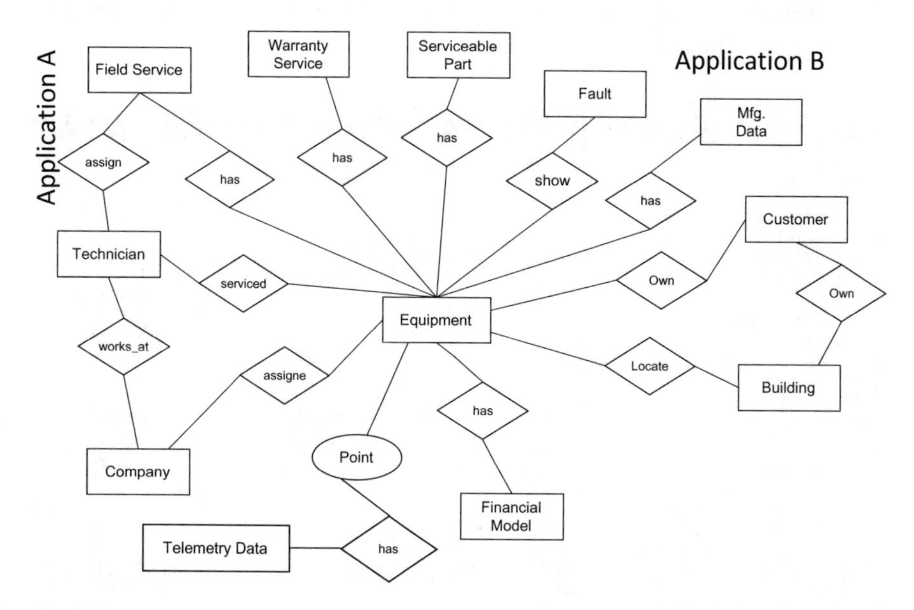

Fig. 7.15 Master data service and its relationship with a connected data model: two applications are easily sharing data and do not require maintaining individual copies of data set

services. A field service staff can follow data link from a report to get service procedures, suggested operating guideline and other product information to maximize a service operation efficiency. Field service data are also linked to operational data (e.g., telemetry data) too. An example of the connected data model is shown in Fig. 7.3b.

Data streams originating in sensors and devices are sent from their point of origin to the Cloud (IoT hub or Kafka) and are then processed in a streaming manner [4, 20]. Let us take the example of a Connected product installed at a customer location. Sensors on this product collect the states of various components of the product. All these states are packed into messages and sent to the Cloud IoT hub on a periodic basis. Once these messages are received by the Time Series API, they are unpacked and immediately pushed into the Storage service. Each message is then examined to determine the additional data that may be required to process the message. This additional data may be in the form of metadata (e.g. Product type, age, etc.), historical content tags (prior incidents of faults, service history, etc.) as well as the definitions of data aggregation and transformation operations that need to be performed on the data for generating analytics (cleansing, filling, aggregations, windowing operations, etc.) The additional data is then fetched from the Store (typically from the in-memory cache), and the message is enriched.

This message is then sent to the Processing service. The processing service performs the required operations and generates metrics (transformed time series data) and analytics (e.g. a tag that the product may be at a high risk for a safety shutdown in the next 24 h). These metrics and analytics are then pushed back into the storage service.

Applications built on top of this data platform queries to the Time series API. These queries are typically served directly from the storage service ensuring low round-trip time (RTT) (Fig. 7.16).

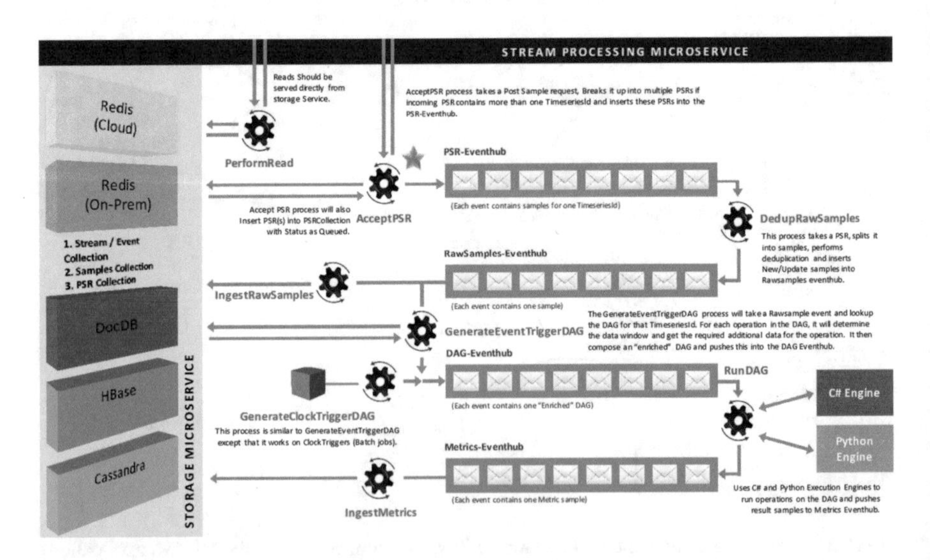

Fig. 7.16 A multi-modal IoT data management and stream processing architecture

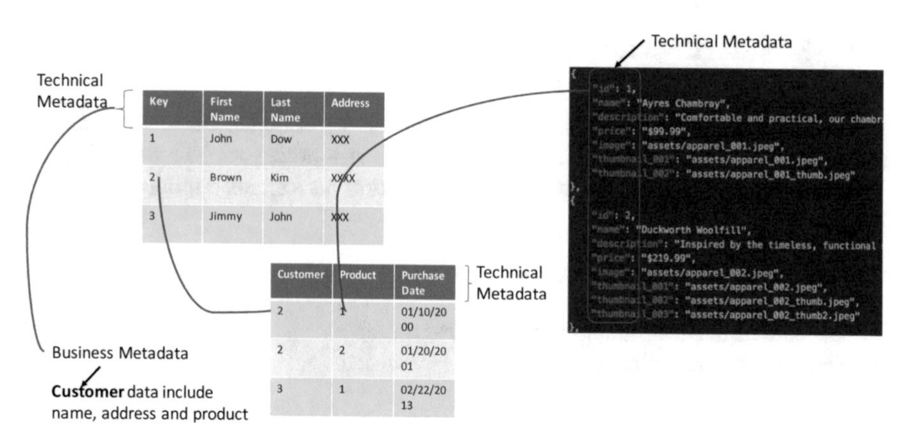

Fig. 7.17 Business metadata vs. technical metadata

7.5.3 *Modeling and Managing Metadata*

As some data sources and a size of data increases, metadata to contextualize acquired and analyzed data in IoT data service platform is highly important. Metadata is about defining data context by defining and answering the followings:

- Who is a creator, governor, user, owner, and regulator?
- What is the business definition, rules, security level, naming standards?
- Where is stored, originated, used, backed up, regulated?
- Why are we storing?
- When was created, updated, and deleted?
- How is this data formatted, and many data stores?

Metadata exists in many information sources across and beyond organization such as social media, open data or purchased external data. There are two types of metadata; namely business metadata and technical metadata. Often Business metadata exists as terms and definitions, while technical metadata represented as a data structure or a schema. In IoT data service, metadata must be shared across all applications and services. Also, all stakeholders including developer, business owner, data architect, auditor, database architect must be able to view them through reports or other delivery tools. Metadata sharing will minimize ambiguity of data definition, simplify data schema design, improve productivity in business process execution through visualizing lineage. Data models, column names, and database schema are a rich source of metadata that may be table names, record names, etc. Figure 7.17 illustrates examples of business and technical metadata and their relationships.

7.6 Monitoring and Managing Data Platform

Managing and monitoring data service platform to meet higher SLA requires a set of toolchains and DevOps with well-defined business processes. The following are minimal features required when you select platform performance monitoring.

- Real-Time application performance
- Performance history and trend analysis
- Network traceability
- The line of business and customer segmentation
- The line of application segmentation
- Custom view
- Self-service view
- Application service level management
- Unified management of across technologies (Spark, Hive, Storm etc.)
- Task performance monitoring
- Notification of performance degradation and failure
- Usage report per application, customer, or line of business
- Health report of cluster and node
- Cluster deployment

7.7 Summary

In this chapter, we have reviewed various data management and processing challenges in developing IoT data service platform and have presented a reference architecture. An IoT data service platform must deal with heterogeneous data types and models. Varieties in storage technologies and data models make ACID management complex, but multi-model storage and data abstraction layer with a separate ACID management service will be an option for make IoT data management simpler and ease for developers.

We have discussed various data ingestion methods including batch loading and real-time ingestion with data flow [21] is driven tool chains including NiFi, Flume, and Gobblin.

We briefly touched a topic related to metadata management that is one of the critical elements in IoT data service platform.

References

1. JCI, http://www.johnsoncontrols.com/buildings/specialty-pages/connected-chillers
2. (Tamara Dull, 2015) Tamara Dull, http://www.kdnuggets.com/2015/09/data-lake-vs-data-warehouse-key-differences.html
3. (JCI DM API, 2016) Johnson Controls' Data Platform API, retrieved from https://my.jci.com/sites/PlatformEngineering/DataDocs/Forms/AllItems.aspx

4. (Fernandez , 2015) Raul Castro Fernandez, Peter Pietzuch, Jay Kreps, Neha Narkhede, et al. Liquid: Unifying nearline and offline big data integration. In 7th Biennial Conference on Innovative Data Systems Research (CIDR), January 2015.
5. (Fowler, 2011) Martin Fowler, "Polyglot Persistence," http://martinfowler.com/bliki/PolyglotPersistence.html
6. (Flume, 2016) Retrieved from http://flume.apache.org/index.html
7. (Thrift, 2016) Retrieved from https://thrift.apache.org/docs/concepts
8. (Chukwa, 2016) Retrieved from http://chukwa.apache.org/
9. (Gobblin, 2016) https://gobblin.readthedocs.io/en/latest/
10. (NiFi, 2016) Apache NiFi, retrieved from https://nifi.apache.org/docs/nifi-docs/html/expression-language-guide.html)
11. (Sqoop, 2016) Retrieved from https://sqoop.apache.org/docs/1.4.6/SqoopUserGuide.html
12. (Grover, 2015) Mark Grover, Ted Malaska, Jonathan Seidman, Gwen Shapira, "Hadoop Application Architectures," O'Reilly, 2015.
13. (Brumbach, 2003) Brumbach, Industrial Maintenance-Text Only – 2nd edition ISBN13: 978-1133131199, Morgan Kaufmann.
14. (Dean, 2004) Jeffrey Dean and Sanjay Ghemawat, https://static.googleusercontent.com/media/research.google.com/en//archive/mapreduce-osdi04.pdf
15. (Perera, 2015) Srinath Perera; Frequent Pattern Mining, Realtime Analytics, Streaming Analytics, Retrieved from http://www.kdnuggets.com/2015/08/patterns-streaming-realtime-analytics.html
16. (Dolas, 2016) Sheetal Dolas, 2016, "Design Patterns for Real Time Streaming Data Analytics, https://www.safaribooksonline.com/library/view/strata-hadoop/9781491924143/part93.html
17. (Brick, 2016) https://brickschema.org/
18. (Haystack, 2016) Retrieved from http://project-haystack.org/
19. (PADIS, 2016) Passenger and Airport Data Interchange Standards (PADIS) Board http://www.iata.org/whatwedo/workgroups/Pages/padis.aspx
20. (MS Azure IoT, 2015) Azure IoT Hub, https://docs.microsoft.com/en-us/azure/iot-hub/iot-hub-what-is-iot-hub
21. (Akidau, 2015) Tyler Akidau, Robert Bradshaw, Craig Chambers, Slava Chernyak, et al. The data flow model: A practical approach to balancing correctness, latency, and cost in massive-scale, unbounded, out-of-order data processing. Proceedings of the VLDB Endowment, 8(12):1792–1803, August 2015.

Chapter 8
Performing Analytics

In the previous chapters, we have introduced various ways to collect and store data from connected devices. Data Analytics is the final frontier for generating the insights from gathered data. The goals of analytics are discovering useful information, suggesting conclusions, and supporting decision-making through a series processes including inspecting, cleaning, transforming, and modeling data. There are many different analytical techniques which can help generate insights. From the traditional business intelligence to extensions like massively parallel processing, to more sophisticated approaches like time series analysis, machine learning and advanced data mining, you can use all of them today because the power of IoT and cloud computing allows you to do so. These are very contextual to the type of data you have, and what you are trying to achieve with the data. There is no silver bullet with a prescriptive list of bests in class analytical techniques by industry domain or use case. Your best approach is to have a multi-model approach to building your analytics allowing you greater flexibility and capabilities. Building a winning data science team is also important to derive success, too. To make them more impactful on a longer term, some of your analytical capabilities have to be self-learning and adaptive. In this chapter, we shall introduce you to the world of possibilities around analytics and how to best leverage them for your desired business outcome.

We shall cover the following topics in this chapter:

- Different Types of Data and Analytic Approaches
- Quantitative vs. Qualitative methods
- Descriptive vs. Predictive vs. Prescriptive
- Example of predictive maintenance in service industry
- Merging domain and data context for building powerful analytics
- Understanding why different analytical techniques are required
- Building your analytical platform
- Tools and programming
- Maintaining your analytical effectiveness

S.R. Sinha, Y. Park, *Building an Effective IoT Ecosystem for Your Business*,
DOI 10.1007/978-3-319-57391-5_8

8.1 Different Types of Data and Analytic Approaches

8.1.1 Qualitative vs. Quantitative

Qualitative. Analytics is used to gain an understanding of underlying reasons, opinions, and motivations. It is non-statistical; its methodological approach is primarily guided by the concrete material at hand. It is information about qualities; information that can't be measured. Qualitative Research is also used to uncover trends in thought and opinions, and dive deeper into the problem. Qualitative data collection methods vary using unstructured or semi-structured [1, 2].

Quantitative. Analytics is used to quantify the problem by way of generating numerical data or data that can be transformed into useable statistics. Some examples of quantitative data are income, height, weight, and the length of arms. It is considered to have as its primary purpose the quantification of data. That allows generalizations of results from a sample of an entire population of interest and the measurement of the incidence of various views and opinions in a given sample. It is used to quantify attitudes, beliefs, behaviors, and other defined variables—and generalize results from a larger sample population. In quantitative analytic approach, the sole approach to data is statistical and algorithmic and takes places in the form of tabulations. It gives insights into the problem to develop ideas or hypotheses for potential quantitative research. Most of the traditional data analysis techniques belong to this approach.

Table 8.1 summarizes the difference between them.

8.1.2 Descriptive vs. Predictive vs. Prescriptive

Descriptive Analytics: These are analytics that describes the history based on accumulated data. Descriptive analytics are useful to explain and to learn from past behaviors. They provide methods for understanding how the past events might influence future outcomes of the statistics we use fall into this category. Some of the

Table 8.1 Comparisons of qualitative vs. quantitative data analysis

Quantitative		Qualitative
Test hypotheses, answer targeted questions	Purpose	Discover ideas or general ideas
Measure and test	Research approach	Observe and interpret
Structured responses	Data types	Unstructured responses
Results are objective	Researcher independence	Results are subjective
Large/big	Sample size	Small
How many visitors per day?	Example questions	What motivate a target group visit the web site

business-relevant questions in this category are "where are products are located by US State?", "Top-K products and services are sold together?"

Predictive Analytics: These analytics are about understanding the future. Predictive analytics provides us with actionable insights based on data. They provide a quantitative likelihood of a future outcome based on observations. One of the well-known applications most people are familiar with is the use of predictive analytics to produce a credit score.

Predictive analytics require a vast amount of data that you have and fill in the missing data with best guesses or with external sources. They combine historical data found in ERP, CRM, HR and POS systems to identify patterns in the data and apply statistical models, and machine learning algorithms to capture relationships between various data sets. Predictive analytics can be used in many organizations and industries including finance, sales, marketing, service, banking, manufacturing, etc. They can help forecast demand for inputs from the supply chain, operations, and inventory and predict failures of assets in-service equipment.

Prescriptive Analytics: These types of analytics are all about providing advice. Prescriptive analytics attempt to quantify the effect of future decisions to advise on possible outcomes before the decisions are made. Prescriptive analytics suggest not only "what will happen," but also "why it will happen." These are recommendations that will take advantage of the predictions. Prescriptive analytics are relatively complex and always comes with a risk of selecting or accepting prescriptions. Many e-commerce companies are successfully using prescriptive analytics to optimize production, pricing, scheduling and inventory in the supply chain optimization and to increase customer experiences (e.g., to make sure that they are delivering the right products at the right time, the right price). Another example is real-time dynamic risk mapping of drivers based on road condition and driver's behavior to create optimized insurance pricing and bundle (http://www.informationweek.com/big-data/how-allstate-uses-mobile-data-to-enhance-customer-experience/a/d-id/1321949). Netflix's "House of Cards" is another example of how data-driven approach is used in content (i.e., product feature) development to forecast "what viewers like to see." Figure 8.1 illustrates a summary of different questions to be answered with various analytic techniques.

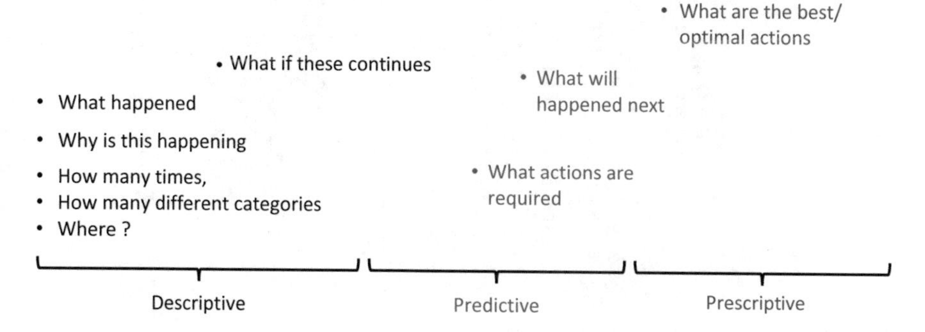

Fig. 8.1 Comparisons among descriptive vs. predictive vs. prescriptive analytics

8.2 Example of Predictive Maintenance in Service Industry

One of the popular use cases of data analytics in IoT domain is predictive maintenance. It is designed to help determine the condition of in-service equipment to predict when maintenance should be performed. It predicts where, when, and why asset failures are likely to occur and help to quickly identify a cause of a failure or variables as part of the causality analysis process. Figure 8.3 illustrates typical steps involved in the field service repair and maintenance. Figure 8.2 shows a reference architecture of predictive maintenance that uses many data collected from external data, IT and OT systems. Prediction can be made at a system level and or a component level. That is entirely up to data availability. If we don't have enough observations at component level failures and operational information, we could omit the component level prediction.

Many different types of analytics must be applied to optimize overall operations. For example, fault detection must be implemented to minimize time to detect failure, diagnostic and prognostic analytics should be applied to make a fast decision in identifying the cause of failure.

Depends on the prognostic outcomes, repair parts must be ordered and delivered, where inventory and supply chain need to be optimized to achieve the maximum outcome from a holistic service operation perspective. In the next few sections, we will discuss tools and technologies for achieving optimal outcomes. There are four steps in developing the solution; namely (1) data gathering and feature engineering, (2) development of a predictive model, (3) deployment of a model, (4) continuous improvement.

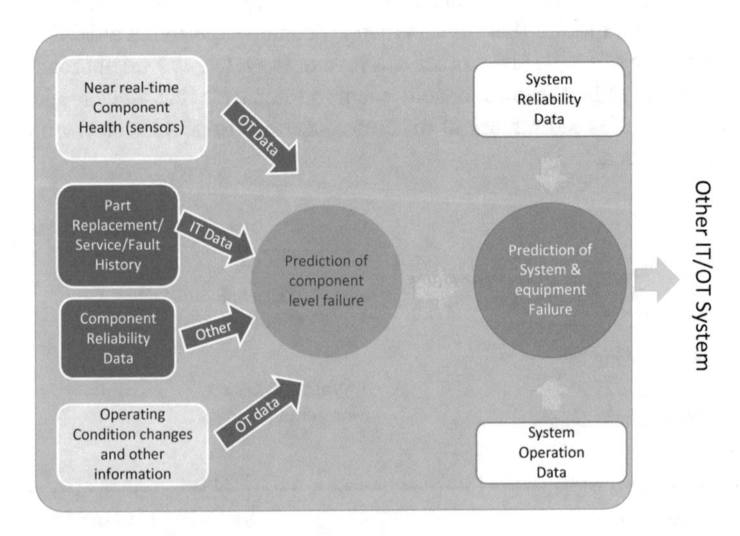

Fig. 8.2 Data fusion approach in Predictive maintenance

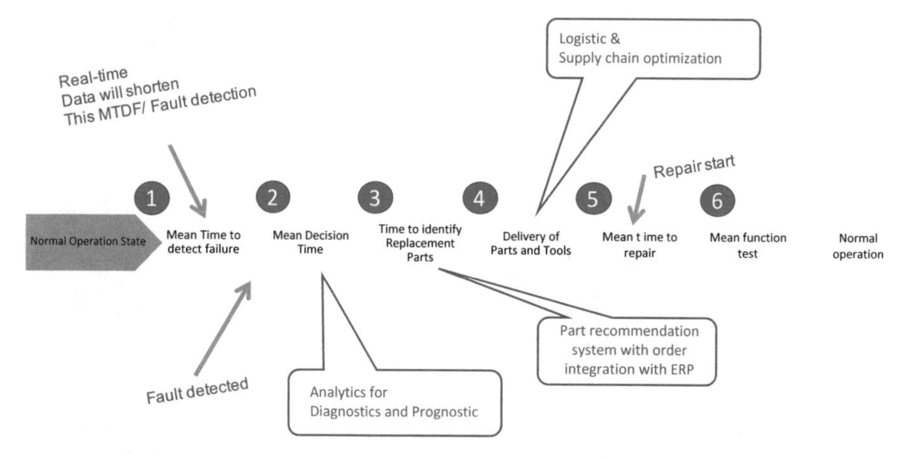

Fig. 8.3 Various stages in maintenance and repair of in-service equipment

8.2.1 Data Gathering and Feature Engineering

To build predictive models, we gather basic product information and historical data to build and train the statistical or machine learning model which can then recognize hidden patterns and further identify these patterns in the future data. It is important that the model captures the relationship between features and the target of prediction, where domain experts can help feature engineering.

Basic Product Information: manufactured or installed/in-service year information along with operational specification (e.g., PSI of a compressor) to create a base model of equipment to be analyzed.

Failure, Maintenance, and Repair History: An essential source of data for predictive maintenance solutions is from the maintenance history of the asset. At minimum, service history of an asset should include the followings:

- asset details
- customer details
- the failure types
- time and date associated with the service event
- the component replaced
- Is planned scheduled services
- Is unplanned emergency repair
- Corrective actions

Maintenance history of the asset enables us to build an initial predictive model. It is imperative to capture these events as these affect the degradation patterns and absence of this information causes misleading results. Few examples of such events in mechanical system maintenance industry are as follows:

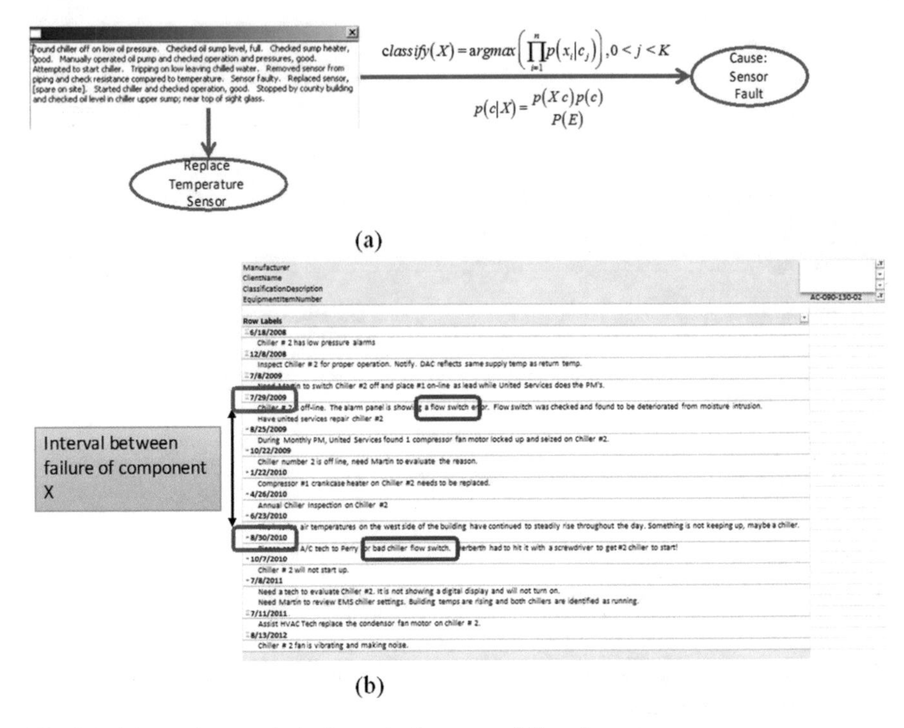

Fig. 8.4 Advanced text analytics is required to extract failure data

- Corrosion inspection (effective control of moisture has high correlation).
- Motor failure, fan, valve failures (purge float, condenser, economizer float).
- Copper chloride deposits, Tube fouling, strainer clogged, control panel failure, relay failure, and excessive throttling of compressor. Often, operating behaviors and conditions are also contributing failures.

Failure events are very rare. However, the learning algorithm needs them to model during training. Often, we need to model both the normal operation pattern and the failure pattern. Hence, it is essential that the training data contains a sufficient number of examples in both categories to learn these two different patterns. Obtaining enough failure events becomes a science in practices.

It is also common that failure data exist as a text or note in a field service application. As shown in Fig. 8.4, extracting a true root cause of failure requires advanced text/natural language understanding, more specifically, text clustering or text summarization. Furthermore, we may have to reference service part transaction data from the different data source (e.g., part sales management system) to get actual failed components. That involves a development of domain specific dictionary that should include repair part number, product name, domain-specific terminology, etc. Often, service note or description are conventional, very noisy and real failure may not directly be observable from them. In fact, this failure data extraction is one of complex machine learning problems.

Once we obtain failure data, we need to bring relevant maintenance history and operational data along with positive examples (e.g., non-failure data).

Operating History: In conjunction with failure and service history from IT systems, we must have operational data to predict how long a machine will last before it fails. We assume the machine's health status degrades over time during its operation and operating conditions. Therefore, we must capture data to contain time-varying features that exhibit an aging pattern and any anomalies that lead to degradation (i.e., telemetry data stored in long-term historian). In IoT applications, the telemetry data from different sensors should capture to detect degrading signature (can be a series of events) during this time frame before the actual failure. Some of the physical and environmental sensing examples related to mechanical systems failures are as follows:

humidity, temperature, the frequency of start/stop, run/operating hours, daily/weekly/monthly variation of load, quality of maintenance program, the pressure of oil pump, contact oxidation (visual inspection), low contact pressure on an auxiliary relay, starting circuit, etc.

Training data are comprising of teaching label, operational data, and maintenance event. In Fig. 8.5a illustrates training samples components. The left-hand side portion shows normal samples labled as 0 and the right-hand portion illustrates failure samples for model development.

Another challenge in data preparation is to determine a right interval size for time series analysis window. Multi-dimensional data analysis will add additional complexity in data processing. There are techniques that do not require a fixed size windowing operations such as Bayesian or transformation based feature comparison approaches. Final feature representation is shown in Fig. 8.5b.

8.2.2 Develop a Failure Prediction Model

There are many machine learning and data mining approaches to solving this prediction problem. In this example, we will use a classical classification/prediction method called logistic regression [3] that is the technical term for drawing straight lines or plans through data points. This dividing plane is called a *linear discriminant* [4] because it is linear regarding its function, and it helps the model 'discriminate' between feature vectors belonging to different categories. In above example, we can draw a line that separate fault samples and a normal sample, shown in Fig. 8.6. Drawing a line through the data is a very significant step. Although the line may seem obvious to us, it requires somewhat complicated math operations on the computer. Few benefits of logistic regression are (1) it's fast, (2) it is easy to understand prediction behaviors, (3) it doesn't require input features to be scaled, (3) it doesn't require any tuning, (4) it's easy to regularize, and (5) it reasonably works well with small number of samples.

In logistic regression, we use a special hypothesis class to try to predict the probability that a given an example belongs to the "1" class versus the probability that it belongs to the "0" class.

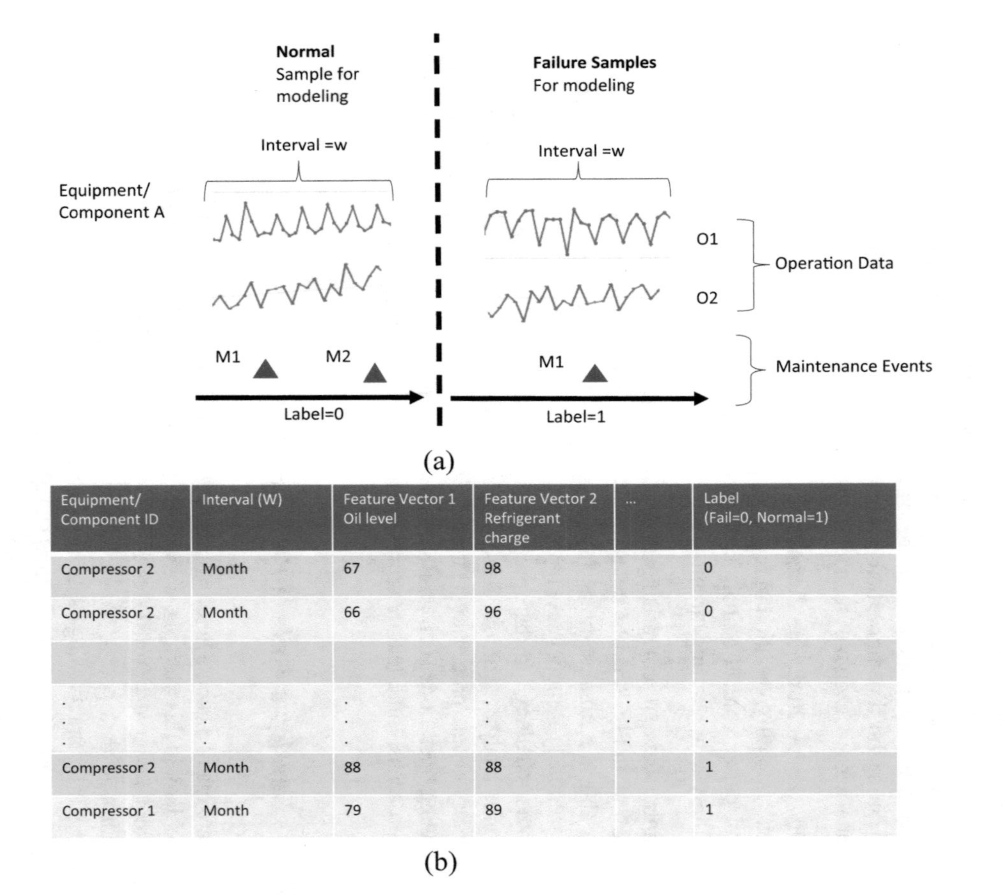

Equipment/ Component ID	Interval (W)	Feature Vector 1 Oil level	Feature Vector 2 Refrigerant charge	...	Label (Fail=0, Normal=1)
Compressor 2	Month	67	98		0
Compressor 2	Month	66	96		0
. . .	. . .	. . .	. . .	. . .	. . .
Compressor 2	Month	88	88		1
Compressor 1	Month	79	89		1

(b)

Fig. 8.5 (**a**) Data preparation for model training. (**b**) The data format of training samples for answer "Will asset X fail in a month?"

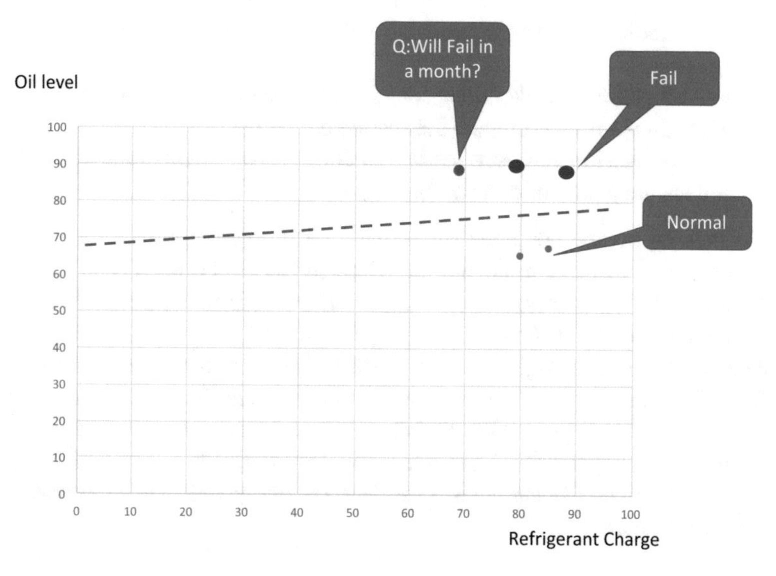

Fig. 8.6 Using logistic regression, we can divide samples into two groups; fail in a month and normal in a month

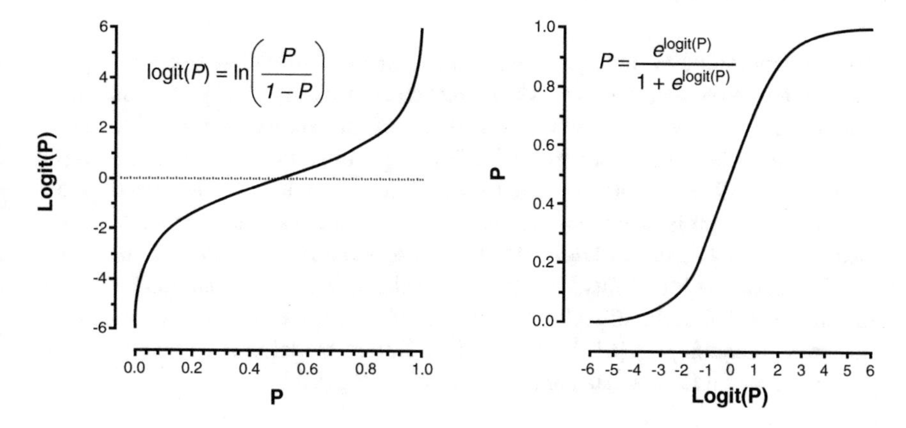

Fig. 8.7 Logit(P) function

In a nutshell, logistic regression [3] is to learn or estimate Logit function that is a simply a log of odds for observations. This function creates a sigmoid (s-shape) curve with probability estimation. Figure 8.7 illustrates the shape of Logit function and its relationship with Probability [0, …, 1]. We are trying to estimate most closet Logit function per given input, that is very similar to the square error in linear regression and can be defined as the likelihood function. The likelihood function is a function of the parameters of a statistical model given data.

A very high-level procedure to get a set of optimal parameters to estimate a Logit function is computed through,

- Given a likelihood function; probability density for the occurrence of a sample configuration
- Compute derivative of likelihood function from log likelihood function
- Use Hessian matrix to compute the second derivative
- Then apply Newton-Raphson method to converge optimal parameter set (i.e., it maximizes the posterior class probability)

The flowing python notebook shows the logistic regression-based prediction of our example and answers to the question: "given Oil level = 69 %, and Refrigerant charge level = 88%, will my compress fail in a month?" In our example, the answered is "the compressor will not fail with 0.98 (close to 1)."

We have presented a logistic regression as a failure prediction algorithm, and we show it works well with sample data set, which has clear linear separable decision surface. However, it's prediction performance will be ceased when a decision surface is no longer hyperplane. There are other algorithms can perform classification and prediction on non-linear decision surface relatively well. We will review them later in this chapter.

8.2.3 *Data Requirements to Build Reliable Models*

"How many failure events are considered as "enough" to train a model?" There is no clear answer to that question as in many machine learning and pattern classification scenarios. In general, quality and quantity of training data matters. If a dataset contains no relevant features to model failure patterns, then developing model is almost impossible. We also need sufficient failure samples to generalized model. The rule of thumb is "the more the failure samples, the better the model is." Often, we have to adapt theoretical reliability prediction methods, to begin with. Examples are "Handbook of Reliability Prediction Procedures for Mechanical Equipment" by [5] and "SR-332 Reliability Prediction Procedure for Electronic Equipment" by Telcordia Technologies [6]. This theoretical failure model will be updated as we collect more real failure data continuous fashion (Fig. 8.8).

8.2.4 *Improving Models and Data Challenges*

In machine learning or model development, residual evaluations (e.g., Cross Validation) do not give an indication of how well the learner will perform when it is asked to make new classification or prediction for data it has not already seen. One way to overcome this problem is not to use the entire data set when training a learner. The most common method is "k-fold cross-validation" which splits the examples randomly into "k" folds. With this, the data set is broken down into k subsets, and the holdout method is repeated k times. Each validation, one of the k

```
In [51]: import pandas as pd
         import matplotlib.pyplot as plt
         %matplotlib inline
         from sklearn.linear_model import LinearRegression
         data = pd.read_csv('/Users/cparkyo/Desktop/lr-sample.csv')
         test_data = pd.read_csv('/Users/cparkyo/Desktop/test.csv')
         linreg = LinearRegression()
         feature_cols = ['OL','RL']
         X = data[feature_cols]
         tX = test_data[feature_cols]

         y = data.CLASS
         linreg.fit(X, y)

         data['Failure_Prediction'] = linreg.predict(X)
         data.head()
         test_data['Failure_Prediction'] = linreg.predict(tX)
         print (data)
         print (test_data)

   OL  RL  CLASS  Failure_Prediction
0  80  65      0           -0.045804
1  85  67      0            0.040290
2  83  66      0           -0.003028
3  88  88      1            0.971042
4  79  89      1            1.020307
5  57  65      0           -0.033369
6  66  67      0            0.050563
   OL  RL  CLASS  Failure_Prediction
0  69  88      1            0.981315
1  55  63      0           -0.121085
```

Fig. 8.8 Example of logistic regression based two-class problem-solving

subsets is used as the test set, and rest training samples are introduced to a training set. A score of validation is the average error across all k trials. The disadvantage of this method is that the training algorithm must be rerun from scratch k times.

Imbalanced data problem: In machine learning, if there are more examples of one class than of the others, the data is said to be imbalanced. Figure 8.9 illustrates failure sample distribution over multiple years and components. As it exhibits, failures are not uniform across classes. Ideally, we would like to have enough representatives of each class in the training data to be able to differentiate between different classes (similar number of samples per class). If one class is less than 10–20% of the data in two class problem, we can say that the data is imbalanced and we call the underrepresented dataset minority class. In our study, most of the failure distributions are falling into highly imbalanced distribution. The performance of most standard learning algorithms is compromised as they aim to minimize the overall error rate. Therefore, conventional global error measurement such as LMS score are not effective to in this situation. Precision, recall, and cost adjusted ROC curves are more useful.

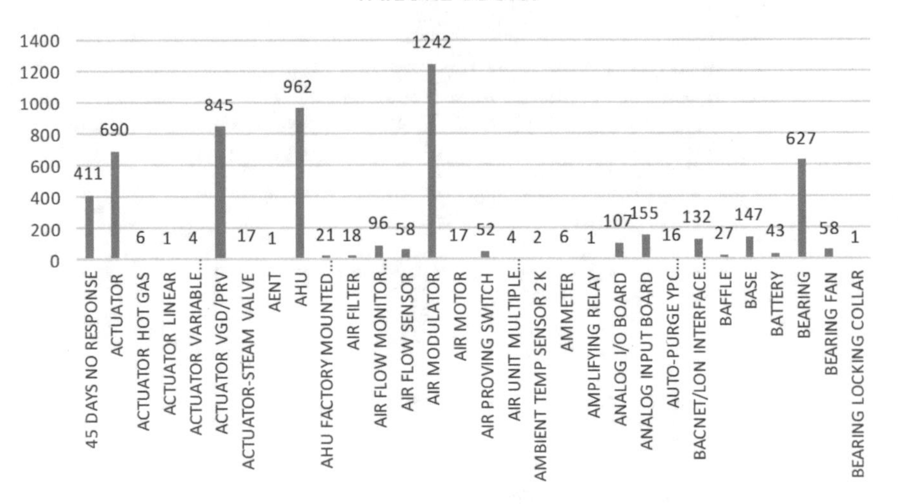

Fig. 8.9 Failures are unevenly occurring over time and different components

8.2.5 Merging Domain and Data Context for Building Powerful Analytics

In our failure prediction modeling, an effectiveness of algorithm is highly relying on data availability and feature modeling (e.g., types of sensors and derived or latent variables that are matters to assets' failure). For example, a proper amount of supply airflow is critical in all kinds of air-based building control systems to maintain desired control effectiveness, energy efficiency and indoor air quality (IAQ) as well as predicting system failures. However, often air flow measurement is not available as physical sensors. Hence, a domain expert can introduce virtual air-flow sensing based on widely installed and available temperature sensors. In conjunction with other sensing data plus virtual sensing, more accurate predictive failure or fault can be modeled and detected. This virtual sensing is an example of how domain knowledge can be incorporated and enhance analytics' performance. Selecting and placing accurate sensing and sampling rates are also critical factors to derive a success of analytics. Again, this is another example of how domain knowledge can improve the performance analytics.

8.3 Understanding Why Different Analytical Techniques Are Required

In the previous section, we have presented a logistic regression-based predictive analysis that works well with linear decision surfaces with labeled training samples. However, in many cases, we may not have training data set available, and decision

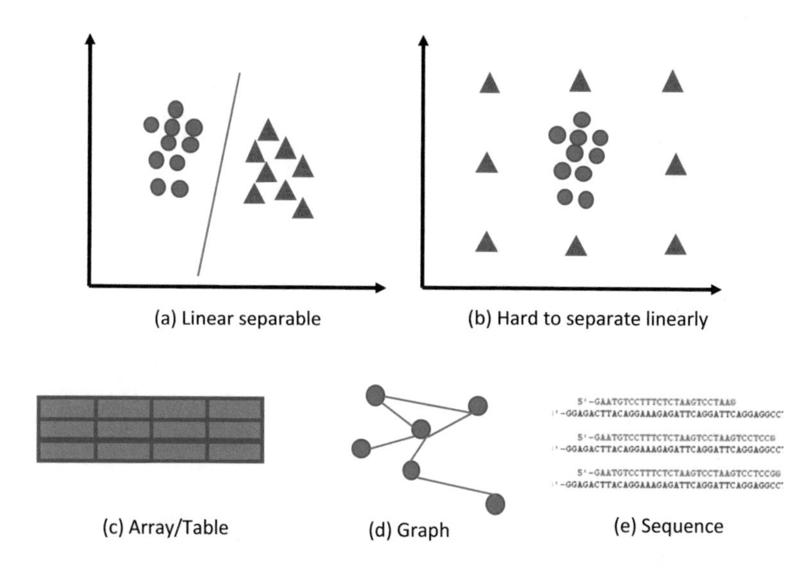

Fig. 8.10 Various data configuration and problems requires different analytic techniques. (**a**) Linear separable, (**b**) Hard to separate linearly, (**c**) Array/table, (**d**) Graph, and (**e**) Sequence

surface is no longer linear as shown in Fig. 8.10b, where unsupervised clustering could be a good option to study insight of data [7, 8].

When decision surface is very complex, and an objective of analytics is to classify or recognize something such as speech and image, then SVM [9] could be an option to try out. When a large training set is available, Deep learning (e.g., higher order neural networks) will be a natural choice [10]. Decision tree [11] and Probabilistic Graph Models [12] are good candidates when a transparency of decision process (e.g., ease to understand how a decision has been made) is a critical requirement.

Depends on a problem we want to solve, we may face different data structures and different analytic objectives. DNA sequence analysis [13] shown in Fig. 8.10e requires substring pattern matching known as sequence analysis very different. Discovery of relationships, a structure of network or influence factor has to deal with a graph-like data structure. A different set of analytics techniques is required in graph analysis such as Apriori-based frequent substructure mining algorithms [14], pattern-growth [15], etc.

As we stated, there are many reasons to have different analytics algorithms. Similarly, it is very difficult to compare machine learning algorithms in general regarding robustness and accuracy. However, we can review some of their pros and cons.

Decision trees: commonly, we use the C4.5 algorithm. The benefit of decision tree learning is "ease to explain" the model. They are however susceptible to overfitting [16].

Bayesian Networks [17] have a strong statistical foundation. They are particularly powerful in problems that inferencing is required over incomplete data and simple to implement. It is very fast.

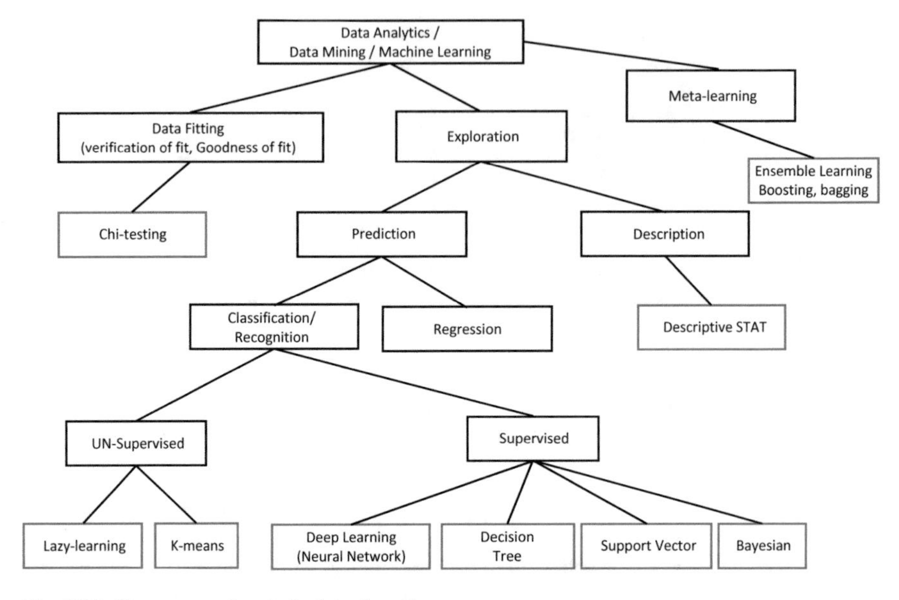

Fig. 8.11 Taxonomy of analytic data algorithms

Artificial Neural Networks are very popular and used in almost every fields of machine perception these days. In theory, they can approximate any arbitrary decision surfaces (e.g., functions). However, they require tuning a large number of parameters (network structure, some nodes, activation functions, etc.). They have the disadvantage in explaining or interpreting model or decision steps and larger computing platform to learn. Incremental learning is very difficult too.

Support vector machine is perhaps considered one of the most powerful techniques. One can always achieve 100% separability with kernel trick. Unlike Artificial Neural Networks, SVM looks to optimize a uniquely solvable problem. They can, however, be computationally intensive and difficult to apply to large datasets.

Ensemble learning techniques are meta-learning algorithms including as bagging, boosting, stacking, etc. Figure 8.11 describe a taxonomy of machine learning and data analytics based on use cases. Again, there is no simple way to say which algorithms are best in performance; rather it is highly dependent on a domain and a problem space.

8.4 Building Your Analytical Platform

In this chapter, we have presented a data platform reference architecture and discussed data processing tools and service. The service provides a run time and job scheduling environment to schedule and executes your analytics written in any language of choice. Your analytics shall access storage service, entity, time series,

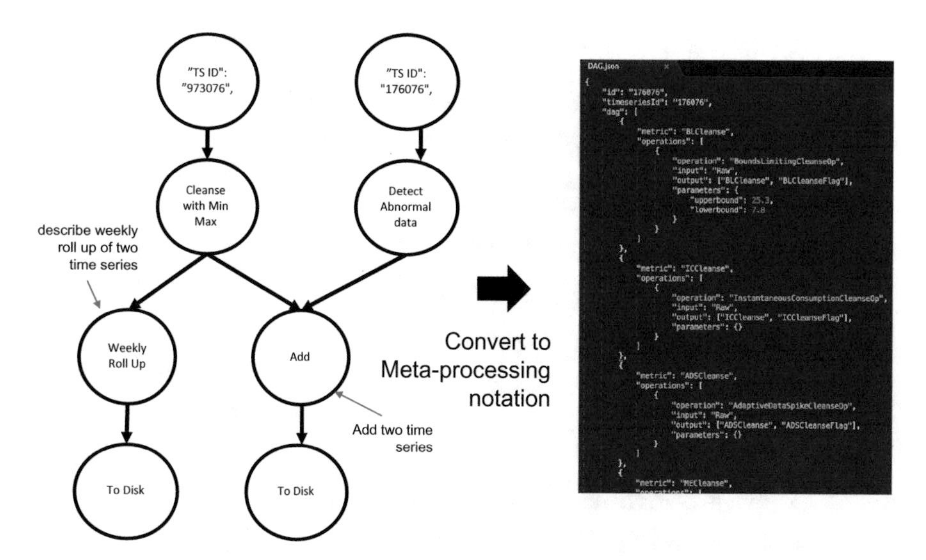

Fig. 8.12 Dataflow driven highly flexible analytic processing definition

historical ERP data via data access interface (e.g., APIs) in model training (see section 7.2 for details).

A developed and tested model can easily be deployed as an analytic job that will produce classification and clustering results. Results of analytics should be available as another data stream. Therefore, data visualization and entity tagging (e.g., add "will fail in a month" to an entity) processes will be accomplished simple and ease. This requires very flexible data flow processing platform. For example, a modern data processing platform provides analytic process flow definition language to describe how analytics will receive input data from and push out to.

Figure 8.12 illustrates an example of analytic job definition graph (Directed Acyclic Graph) that will define the flow of job execution. DAG engine will investigate any duplicate stream processing and dependencies among processes, then optimize the execution.

8.4.1 Tools and Programming

There are too many different commercial and open source tools for data analytics. Virtually any language can be used in developing analytics, but R and Python are popular choices because they have various analytics libraries and existing ecosystems. Per, KDnugget's survey, R remains the leading tool, with 49% share (up from 46.9% in 2015), but Python usage grew faster, and it almost caught up to R with 45.8% share (up from 30.3%). RapidMiner [18] remains the most popular general platform for data mining/data science, with about 33% share. Notable tools

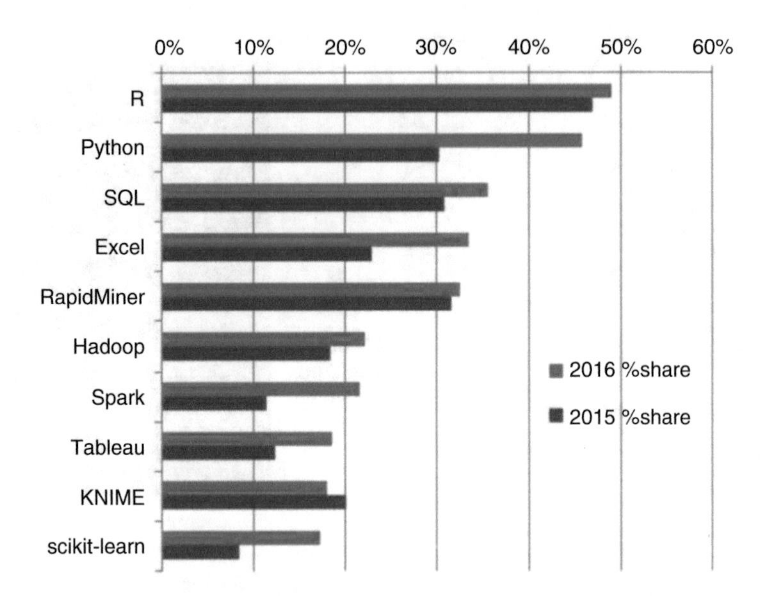

Fig. 8.13 KDnuggets' data analytics/science 2016 Software Pool

with the most growth in popularity include Dato, Dataiku, MLlib, H2O, Amazon Machine Learning, scikit-learn [19], and IBM Watson [20]. Excel and Tableau are popular choices for data visualization (Fig. 8.13).

Deep Learning tools including TensorFlow, Theano, Caffe, and CNTK are getting a lot of attention, however, still in the beginning of its journey. Complete survey result can be found from [20].

8.5 Maintaining Your Analytical Effectiveness

Manu successful analytic platforms and products based on analytics and artificial intelligence including Amazon's Alexa, IBM's Watson, and the recommendation systems of Amazon and Netflix, also known as trained systems, have received big the public attentions. There is also excitement about bringing such capabilities to other applications and various vertical markets in consumer and commercial sector. However, the complexity of building them is very challenging, even for Ph.D.-level computer scientists and engineers. If such systems are to have a truly broad impact, we must make them as "ease of development" and "ease of maintenance." As a result, they can be easily applied to different application domains. We believe the following abstractions will bring complexity in maintaining trained analytical system manageable.

Model and Data Management: Early days, "who has s better algorithm" was matters in evaluating analytic or classification: or recognition accuracy. However, with the advent of machine learning, "who has better and large data and model" is matter and drives the performance and precision of analytic offerings. Now, the analytic offering consists of the data set and models trained with the data set. Hence, versioning of training data set and corresponding model is mandatory and must be a part of a product.

Development Tool and Deployment: To ensure that a trained system is accessible through the application programming interface to developers. API must be must be simple and easy to use. Therefore, many developers can try many algorithms; the ability to integrate diverse data resources and formats requires abstracted entity service APIs to be versatile. A combination of common data models (e.g., canonical data models) and a flexible analytic processing definition language satisfies these criteria. Training data and learned model versioning with processing lineage are becoming critical.

Infrastructure Virtualization: a modern analytic and data-driven product must accommodate many different algorithms on top of big data. It is important to provide a clean integration interface between algorithms and systems to build such products, For example, data virtualization technology completely abstracts underlying storage and data services as well as job executions. Therefore, data analytics team do not worry about and eliminate inter-dependencies between platform and analytics.

8.6 Summary

In this chapter, we have presented various data analytic techniques and tools. A practical example of predictive maintenance has presented with a real-life example. It uses logistic regression as a predictor that can answer "will fail in a month." Effectiveness and weakness of various analytics algorithms have discussed along with various data challenges in building training samples.

Domain expertise is critical to improve and accelerate analytic embedded product development. It can dramatically reduce time in feature engineering and bridge data gaps such as Autoencoders [21, 22], Deep Neural Networks [10].

R, Python, and SQL are still the most popular tools for a data scientist, while Deep Learning is getting momentums. This is the interesting and exciting moment for those who are practicing data analysis. We believe, future is bright and changes rapidly.

References

1. (Wyse, 2011) Susan E. Wyse, 2011, Difference between Qualitative Research vs. Quantitative, retrieved from https://www.snapsurveys.com/blog/what-is-the-difference-between-qualitative-research-and-quantitative-research/
2. (Friedman, 2001) Hastie, Friedman, and Tibshirani, The Elements of Statistical Learning, 2001.
3. (Hosmer, 2013) Hosmer Jr, D. W., Lemeshow, S., & Sturdivant, R. X. (2013). Applied logistic regression (Vol. 398). John Wiley & Sons.

4. (Izenman, 2013). Linear discriminant analysis. In Modern multivariate statistical techniques (pp. 237–280). Springer, New York.
5. (Naval, 2011) Handbook of Reliability Prediction Procedures for Mechanical Equipment http://reliabilityanalyticstoolkit.appspot.com/static/Handbook_of_Reliability_Prediction_Procedures_for_Mechanical_Equipment_NSWC-11.pdf
6. (Pechta, 2002) Pechta, M., Dasa, D., & Ramakrishnanb, A. (2002). The IEEE standards on reliability program and reliability prediction methods for electronic equipment. Microelectronics Reliability, 42(9), 1259–1266
7. (Jain, 1999) A. K. Jain, M. N. Murty, and P. J. Flynn, Data Clustering: A Review, ACM Computing Surveys, 31(3), 1999.
8. (Grabmeier, 2002) J. Grabmeier and A. Rudolph, Techniques of Cluster Algorithms in Data Mining, Data Mining and Knowledge Discovery, 6:303–360, 2002.
9. (Crammer, 2001) K. Crammer, K. and Y. Singer, On the Algorithmic Implementation of Multiclass SVMs, Journal of Machine Learning Research 1. 2001.
10. (Dean, 2012) Jeffrey Dean, et al. "Large scale distributed deep networks." Advances in Neural Information Processing Systems. 2012.
11. (Quinlan, 1986) J. R. Quinlan. Induction of decision trees, Machine Learning, 1:81–106, 1986.
12. (Koller and Friedman, 2009) Daphne Koller, Nir Friedman, Probabilistic Graphical Models, July, MIT Press, 2009, 9780262013192.
13. (Agrawal, 1995) R. Agrawal and R. Srikant, Mining Sequential Patterns, International Conference on Data Engineering (ICDE'95), 1995.
14. (Srikant, 1996) R. Srikant and R. Agrawal. Mining Quantitative Association Rules in Large Relational Tables, SIGMOD 96.
15. (Pei, 2004) Ian Pei, Jiawei Han, B. Mortazavi-Asl, Jianyong Wang, H. Pinto, Qiming Chen, U. Dayal, Mei-Chun Hsu Mining sequential patterns by pattern-growth: the PrefixSpan approach, IEEE Transactions on Knowledge and Data Engineering, Issue 11 Nov. 2004.
16. (Aalst, 2010) Van Der Aalst, W. M., Rubin, V., Verbeek, H. M., van Dongen, B. F., Kindler, E., & Günther, C. W. (2010). Process mining: a two-step approach to balance between underfitting and overfitting. Software and Systems Modeling, 9(1), 87–111.
17. (Geiger, 1997) Friedman, N., Geiger, D., & Goldszmidt, M. (1997). Bayesian network classifiers. Machine learning, 29(2–3), 131–163.
18. (Klinkenberg, 2013) Klinkenberg, R. (Ed.). (2013). RapidMiner: Data mining use cases and business analytics applications. Chapman and Hall/CRC.
19. (Pedregosa, 2011) Pedregosa, F., Varoquaux, G., Gramfort, A., Michel, V., Thirion, B., Grisel, O., … & Vanderplas, J. (2011). Scikit-learn: Machine learning in Python. Journal of Machine Learning Research, 12(Oct), 2825–2830.
20. (Gregory Piatetsky, 2016) Gregory Piatetsky, R, Python Duel As Top Analytics, Data Science software – KDnuggets 2016 Software Poll Results.
21. (Pascal, 2010) Vincent, Pascal, et al. "Stacked denoising autoencoders: Learning useful representations in a deep network with a local denoising criterion." The Journal of Machine Learning Research 11, 2010: 3371–3408.
22. (Minmin, 2012) Chen, Minmin, et al. "Marginalized denoising autoencoders for domain adaptation." arXiv preprint arXiv:1206.4683, 2012.

Chapter 9
Closing the Loop for Value Capture

"Intelligence is the source of technology. If we can use technology to improve intelligence that closes the loop and potentially creates a positive feedback cycle."

—Eliezer Yudkowsky

When we look across the two spectrums of IoT—consumer and industrial, we see different things evolving. On the consumer front we see more focus on distribution of new capabilities to mass markets; on the industrial IoT front, we see more focus on deeper analytics. In both cases, if we do not have enough focus on closing the loop between sensing, data, analytics, and customer outcomes, the promise of IoT will be underachieved. There are many examples of businesses that have failed to take off despite having cool devices, technology, and analytics because they failed to close the loop with the customer. For large existing businesses making a foray into IoT, closing the loop with internal business operations is a huge challenge.

The first part of closing the loop is an effective delivery ecosystem. Your powerful insights derived from rich data analytics needs a number of different delivery channels to become actionable. In general, we have to support multiple delivery models and paths (e.g., electronically, physical, etc.). You need a robust yet very flexible mechanism which allows you to take the same data and insights, and distribute it through different delivery channels. This is a bit more complicated than having good reporting tools or standard APIs; you need intelligence built into your delivery ecosystem for the system to dynamically decide what information to send through which channels at what frequency or at the occurrence of which events. There is an element of machine learning you need to build in your delivery ecosystem. You should also be able to adapt your visualizations and transport to new or evolving consumption expectations.

Let's say you are the manufacturer of capital-intensive industrial machinery. You have built an IoT business of providing diagnostics services around the equipment leading to better operating cost and improved asset life. If your diagnostics requires services to be performed and you have neither your own service network nor a set of reliable service partners that you can recommend, you have a big missed opportunity.

S.R. Sinha, Y. Park, *Building an Effective IoT Ecosystem for Your Business*,
DOI 10.1007/978-3-319-57391-5_9

The last mile connectivity between monitoring and monetizing is a critical component for deriving value out of your IoT investments. This is the second part of closing the loop we started discussing in the last chapter. The traditional business support systems address the needs of two major stakeholder groups—internal organization and customers; in the IoT world, you have to also account for the success of a number of partners and intermediaries. IoT is bringing many players of various sizes to come into a common marketplace to transact data, insights, and services. You need a more comprehensive and inclusive approach to building your business support systems. Also, every business support transaction has to be in the context of IoT devices and data without being too far removed from layers of aggregation and synthesis. So you need a different vector from your normal practices when dealing with IoT-driven scenarios.

In this chapter, we shall start addressing issues with closing the loop and how to address them. We shall cover the following topics in this chapter:

1. Understanding the needs of an effective delivery ecosystem
2. Defining a framework for business support ecosystem
3. Defining customer success
4. Selecting your tools and target systems
5. Distributing data products electronically
6. Creating systems for your channel partners to transact business
7. Integrating with your enterprise IT systems effectively
8. Engaging your stakeholders through reporting

9.1 Understanding the Needs of an Effective Delivery Ecosystem

An effective delivery ecosystem a key enabler for the success of your IoT business. Building such an ecosystem is complicated. There are multiple components that have to be part of it.

- Business applications
- Data visualization and reporting tools
- Electronic product distribution
- Application services and APIs
- Enterprise integrations

An effective delivery ecosystem should have the following characteristics:

- Provides a 360° view of the business by combining analytical insights with transaction data to create a holistic view of customers, their usage of products and data offerings, and the overall IoT business. Such a view is provided in an integrated manner. The increasingly popular concept of digital twins helps create better 360 views by digitizing every dimension of product information from design to operation.

- Makes insights actionable. While the IoT analytical platform will provide a lot of useful insights about the product's usage, an effective delivery ecosystem needs to provide context to its history and the customer using it. This will allow the insights to be actionable and deliver tangible value. For example an industrial IoT device might show degrading performance, however, if the usage is seasonal, there might be appropriate opportunities to fix the issues. If the service technician is equipped with the correct information about degrading performance, possible windows for fixing the issues, and parts and tools required for the fix, it becomes actionable insight.
- Transports data and insights in real-time or near real-time basis. In the hyper-connected IoT world, batch processing of data and insights once or twice a day is insufficient. This needs to happen at least at near real-time basis to ensure that information is current and relevant.
- Is augmented with learning capabilities. Sometimes the various stakeholders involved in the IoT ecosystem may not be familiar with all the tools and how to use the data. An effective delivery ecosystem should have built-in tools to help users understand the systems and data. This is similar to help menu, only more interactive and adaptive to user's needs.
- Has capabilities to articulate and demonstrate value for end customers. The delivery ecosystem needs to be able to capture value creation from product data analytics and enable the customer touch points to be able to communicate that.

9.2 Defining a Framework for Business Support Ecosystem

The business ecosystem we live in today's IoT dominated environment is different than the ones we were used to previously. Not all the changes are because of IoT, some of them are results of natural evolution of technology, business practices, and globalization. Here are some of the main business practices we see as different:

1. Knowledge is easily available, so consequently, it is not the primary source of power. You can learn about anything in minutes. Where you needed specialized training in a trade or discipline acquired through rigorous school programs, you can now learn the basis by just watching Youtube videos. This has two implications—the entry barrier for anybody providing similar products or services as you is lower than in the past, and your customer will explore other choices including self-performance if their needs are not addressed timely.
2. In the industrial age, the relationship between manufacturers, distributors, sellers and customers used to be a sequential linear relationship, one feeding the next one in line both products and information. Now the relationships are multi-dimensional. Irrespective of how you go to the market, you need to have a direct close relationship with the end customer of your products and services [1] (Figs. 9.1 and 9.2).

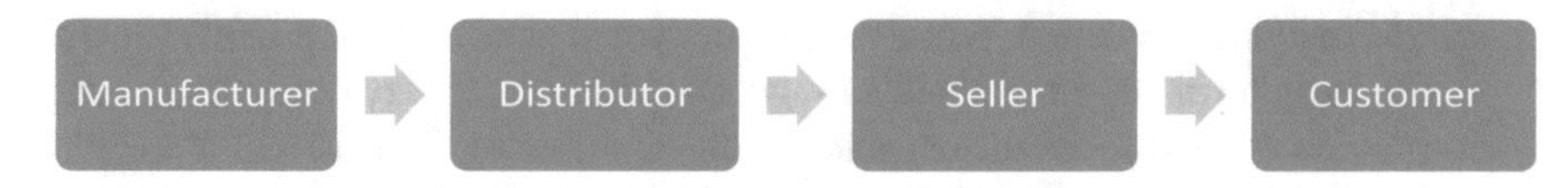

Fig. 9.1 Traditional business ecosystem relationship chain

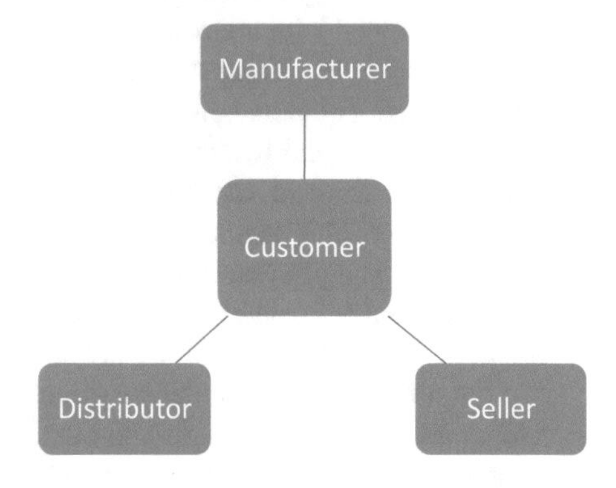

Fig. 9.2 Multi-dimensional business ecosystem

3. Traditionally companies believed in vertical integration from raw material to every customer engagement point, so all functions that produce the good or serve the customers were internal. Now, producers of goods and services have to coexist with a network of independent distributors and service providers. This is to ensure faster scalability in market reach, streamline internal capital structure, and have options to address to changing business needs [2].
4. Long term loyalty to a brand or company is fast getting replaced by alternatives which continuously offer value enhancement for customers. That is how brands like Fitbit and Nest have become celebrities in such a short span of time.
5. Profit margins are shrinking in a global marketplace. So you need to always work on reducing your cost to acquire and serve a customer.

The business support ecosystem has to comprise of many components:

(a) Customer management. In addition to the normal CRM functions like pipeline management, sales process execution, customer support, marketing, and promotions; you need additional extensive customer engagement capabilities including social media or social business, the marketplace for new offerings, a place for customers to identify and select service providers if applicable to your business scenario. This has to be a one-stop solution for all possible interactions between end customers, and your company and offerings.
(b) Partner management. This has to be beyond a portal to register and show partners. Partner management systems should include functionality to manage partner performance, business systems for partners to conduct their business, manage partner engagement with end customers and similar activities.

(c) Commercial management. This includes all functions related to managing billing, receivables, contract management, licensing, etc.
(d) Enterprise planning and execution management. Often as part of your IoT business, you will have to not only interact with the back-end manufacturing and supply chain processes internally but also those of external partners who are part of your offering delivery ecosystem. As part of this component, you need comprehensive functionality for all such transactions.
(e) Business analytics and reporting. Here you will need highly automated, adaptive and intuitive capabilities to get insights into business performance and possibly guide strategies. There should be easy to use tools for various types of internal users to analyze different dimensions of your business.

As part of the business support ecosystem, you also need to enable your partnering strategy. We have covered this topic extensively in Chap. 11.

9.3 Delivering Customer Success

The growth of your IoT business will be largely defined by your ability to deliver customer success. Product or offering is extremely important, but without the enabling ecosystem to deliver customer success, their adoption will be limited. There are a few elements you need to consider when thinking this through:

(a) What is customer's perceived value of success? Due to competitive threats and hyper-evolutionary nature of IoT businesses, you will need very succinct definitions of perceived value expectations by customers. This definition has to be inclusive of expectations from the product and also the lifecycle engagement events customers will engage with the offering and your business. You will need mechanisms for very good definitions and also adapt those definitions based on changing customer preferences and usage.
(b) How do we communicate and validate the value to customers? While it is very important to build the customers' expected value into the offering, it is equally critical to keep communicating the value to customers and get feedback about their experience. This will help address the shortening attention spans and the competitive threat from the proliferation of similar products/offerings. For example, if your customer has brought a solution to help them implement better security at home or reduce energy consumption at work, it is important that you close the loop with them on what was their expectation versus what they have achieved by using your products.
(c) How do we make it easier for the customer to engage with your business when they need help? We all hate when we have to answer endless questions when calling a helpdesk number or wait for a long time to get resolutions or clarifications. One of the best customer service experiences we have seen is from Apple—be it their call centers or in-store tech support [3]. Even if you have to wait for a bit at times, there is some engagement instead of blindly waiting. You must have

multiple methods for customers to seek help or clarify—phone, web, social media, and the product itself. There are a lot of self-learning artificial capabilities available today that can help you make this an intuitive and less expensive proposition. In every interaction, your support system or people have to help customers use the product and get more out of it after addressing the issues. Often customer support centers are oriented for cross-selling, you need to design the support ecosystem to create a pull for your customers to buy more from you.

(d) How do we know that customers' continue to have a positive experience from your offerings and engagement? The traditional feedback mechanisms of forms and surveys are sometimes limited to capture true customer experiences real-time. You need to employ technology like click analytics to capture experiences in moments of interaction. You also need to triangulate multiple sources of feedback in context to get better insights.

Sometimes you need to make organizational or operating model or incentive changes to improve on delivering customer success. Start with changing the organization denotation from customer support to customer success. Whether you are in the consumer or industrial space, this is a key organization which will help with retention and growth. People working in these functions must have excellent engagement skills and deep offering knowledge. They should also have sufficient internal organizational navigation skills to marshal the necessary resources for delivering customer success. The metrics tracked have to shift in their orientation from support to success connotation; the orientation has to change from singularly driving internal operational efficiencies to delivering more customer value.

9.4 Selecting Your Tools and Target Systems

You will have tools for internal usage as well as integrate with external data systems. These extend from business applications to data visualization tools to things like weather data or similar services. These tools and systems are an integral part of your IoT business.

In addition to functional capabilities of the tools and systems in serving the purpose you intend for them, following are some key considerations you should keep in mind when selecting such tools or external data systems:

1. Easy integration with other tools your business might be using to maximize current technology investments and reduce change management issues. If the integrations are achieved using publicly available APIs and in a plug and play model.
2. Easy to use with less training required.
3. Can address the needs of multiple product distribution channels.
4. The functionality and data content is transportable across other tools and platforms. This will help avoid technology or vendor lock-ins.
5. Scalability for different use cases and sizes.

6. Reliability of performance under all circumstances.
7. Security for data and application access.

9.5 Distributing Data Products Electronically

IoT businesses usually have two types of products—physical hardware products and software based data analytics products. There is generally a very close linkage between the two. Typically for any IoT business, there are periodic releases of new data products or updates to existing product capabilities. In this section, we will talk about electronic distribution of data products.

9.5.1 Demonstrating Value of Current Offerings

In order to grow your business in future, first, you need to have a sustained relationship with existing customers and should be able to create demand for your data products. In dealing with this, there are a few key things you need to keep in mind:

(a) If customers currently use non-data products from you, educate them on the value add done by your data products. The value could be added through better visibility of the operational performance of devices or future forecast of something. For example, your customers have brought smart thermostats which manage the heating and cooling of homes or workplaces. Now you have a new offering which helps you lower customers' energy bills through analytics and simulation.
(b) Track the usage of your current data products purchased by customers. If you see the lack of usage, you need to understand why and take actions on remedial measures. Continuing the same example from above, your offering includes energy usage dashboards on the device. You want customers to see that dashboard and read through the interpretations as an indicator of their interest and awareness. Collecting product usage through click analytics and other similar methods is a good way to gauge usage and interest. You may find insights into what features or capabilities customers are most interested in; that should be an input for your future product design.
(c) Constantly demonstrate the value of your data products that your customers are using. If they do not see the value, their propensity to buy anything in future will be low. You can do this by usage tracking and direct feedback. Using the same example, you should intimate your customer how much energy and cost savings ideas has the smart thermostat identified, how many of them have been implemented and what is the net residual value creation.
(d) Initiate training whenever you see the lack of appropriate adoption and usage of your current offerings.

(e) Keep reminding your customers about how your offering has evolved to create more value through the length of the relationship. This makes the customer part of your journey and demonstrates how you both are growing.

All of the above steps help you build trust with the customer and gives you permission to sell for future.

9.5.2 Positioning New Offerings

As you release new data products, you need to take it to your existing and new customers. For existing customers, you already know or have the ability to know a lot about them. It is critical that you understand the customer profile, behaviors, preferences, and use that to put a context to the perceived value of the new offering specifically for your customer. This has to be as personalized as much as possible. This has been done very well in the online advertising world. Try searching for Petra, Jordan once from the browser of one of your devices, you will keep getting tour offer displays in the ad pane, Facebook posts and various other marketing inserts from different providers. On similar lines, let's say you have an IoT activity tracker. You can use the customer activity data to suggest on hand buying activity enabling tools like shoes, clothes, etc. and on the other hands more apps which help with different aspects of activity analytics. In an industrial context, if you have an IoT solution comprising of multiple sensors and analytics on the sensor data, you can position new diagnostic services to the customer around things for which there may not be sensors currently.

For new customers, where you may not have enough background information, it is important to position your offering in the context of the segmentation to which your targeted customer belongs. In the consumer segment this will largely be driven by demographic analysis and in the industrial IoT context, it will be influenced by industry vertical and user type profile. For example, you may position a range of energy management solutions to facility operators and a different range of predictive analytical offerings to facility service professionals.

9.5.3 Implementing Monetization Infrastructure

Monetization can be implemented through multiple licensing models like perpetual, subscription based, utility based, or value based. Each of these models requires an enabling infrastructure.

There are several components which make up a robust monetization infrastructure.

- Customer relationship management (CRM). This is important to understand your customer profile, background, needs and what offerings they currently consume. There should be capabilities to offer the customer trial or demo experiences of offerings they might be interested in. This should have a linkage to the customer ordering process but a lightweight one. Many companies will allow prospective customers to simply download an offering with limited term or capability or both. A good CRM system should have options for managing offerings portfolio including relevant collateral, recommending new offerings, managing sales campaigns, and tracking various sales events.

- Customer order management (COM). COM takes the inputs from CRM when the customer makes the purchase decision and drives it through the fulfillment system. Typically every business has an order management system which is tied to other financial systems.
- Fulfilment. This includes multiple things:
  - Entitlement determines what software or data output the customer should have access to depending on their purchase decision
  - Content delivery takes the software or data components residing on various storage or hosting locations based on entitlement
  - Software packaging assembles the right software and data kernels from content delivery system for usage by the customer as per entitlement
  - Licensing applies the appropriate rights for software or data usage to the customer
- Installation. The software or data packet needs to be transported from the content delivery system to the customer's device.
- Activation. Based on licensing, the use of data product needs to be authorized at the device for the duration of the contracted use.
- Lifecycle management. This covers every dimension of the life of the data product or software including upgrades, obsolescence, discontinuation, feedback, problem reporting, and resolution.

The Fig. 9.3 is a pictorial representation of how these components need to work in tandem to realize your monetization objectives:

9.6 Creating Systems for Your Channel Partners to Transact Business

As we have touched upon earlier in this chapter, it is an imperative of IoT businesses to have enabling systems for your partners to transact their business within your ecosystem. This is important for better retention of partners and influence over them. Doing so will help you from maximizing your opportunities with the end customers, because you may have to rely on partners for components of your IoT offerings.

An IoT-enabled business operating system integrates real-time connected device health monitoring, failure based product or service part recommendation service,

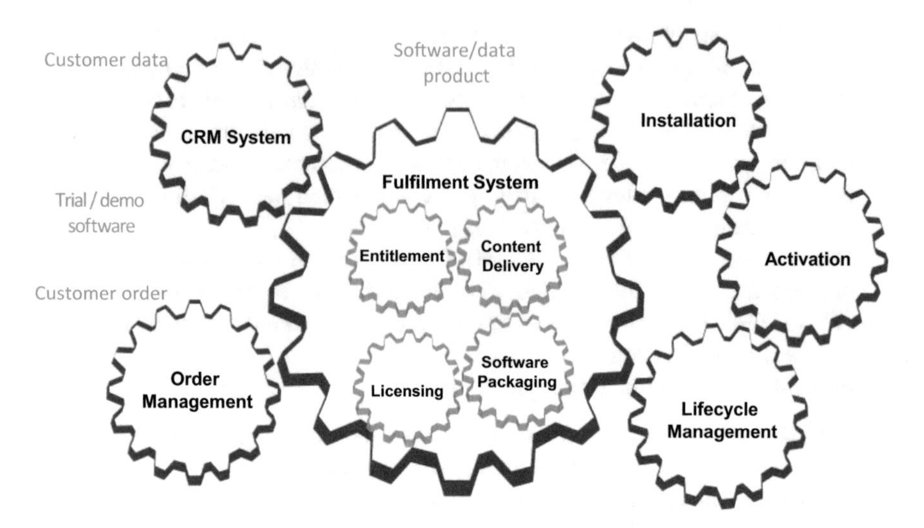

Fig. 9.3 Monetization Infrastructure

service visit scheduling and finally customer relationship management. This integrated solution provides an environment to complete all necessary business transactions highly integrated and automated matter for a service provider. It enables multi-party communications including a connected device, an OEM, a service provider and a requester.

When creating systems for your channel partners to transact business within your ecosystem, consider the following:

(a) Construct a single place to work with customer and opportunity information and that should be linked to how the opportunity was arrived at. For example, if there is a repair opportunity identified through data based diagnostic techniques, the relevant data should also be made available along with associated performance data.
(b) Build a process to rate and identify the best channel partners. You may want to develop your proprietary algorithm to do so as a competitive differentiator to attract new channel partners and provide better customer assurance.
(c) Create a marketplace for customers and channel partners to transact business without requiring any other third party system.
(d) Design a way to capture value from transactions like a new product or offering sales, % of revenue, etc.
(e) Perform analytics on identifying white spaces to drive further channel development.
(f) Construct a platform for launching new offerings.
(g) Support learning tools to help channel partners get training on your offerings, core technologies, and customer support.
(h) Enable social business and connected communities.

Security is always a big concern when you have systems with third party access. The Target breach of 2013 brought this to the forefront of every conversation. You need to build robust policies and procedures for both remote access to your business systems as well as access by third party people and systems. For IoT businesses implementing robust protocols is critical because there might be access to IoT devices by these third party resources and systems which in turn can become the gateway to other potential breaches. It is better to restrict all third party interactions to separate firewall areas or DMZs. Any intrusion should immediately invoke stringent segregation protocols to limit threats. You also need good legal and contractual protections for such eventualities to limit your liabilities if there is any issue because of your partners. Explore options to safeguard you through insurance as well.

9.7 Integrating with Your Enterprise IT Systems Effectively

Many companies have made great strides in building out their IoT platforms and offerings. Sometimes they fall behind in integrating these new businesses appropriately with their internal enterprise IT systems. Such gap limits the full potential of opportunities generated through IoT. For example if you have IoT-enabled capabilities to do predictive diagnostics and identify opportunities for additional services, but do not have a way to transfer that opportunity to the CRM or opportunity management system automatically for tracking and closure, it may fall through the cracks and not get realized.

Integrating with internal IT systems can be difficult due to many reasons:

(a) Organizational silos and lack of collaboration between your IoT business and IT organization
(b) Lack of technical maturity and sophistication in your internal IT systems
(c) Insufficient funding for infrastructure and process upgrades

But seamless integration is key to your long-term success and full potential value capture. Here are some best practices from our experience you can pursue:

1. Educate your enterprise IT team on the objectives, business model, technology architecture and integration expectations from your IoT business. Your IT team wants to be part of your winning game, they want to contribute, but you need to take the first step in bringing them onboard. Upon education, you need to align the agendas of various groups into a common theme targeted towards growth.
2. Define a common semantic model to interpret customer, product and transaction data. This is a challenge for many companies because the evolution of IT systems and IoT businesses being on different timelines might have different definitions or interpretations of customer, product and transaction data. Multiple disjointed systems often accentuate the problem, for example, a part number for a product or offering could be different in various systems. In rare cases, the relationship between a customer, product, and business transaction is non-linear; there might

be different buyers for different services at different points in time for the same base installed product creating a complicated relationship.
3. Create a unified data lake for operational data from products, customers, and internal IT systems. Unless there is synergy between the platforms which has product operational data analytics and internal IT data analytics, it will be hard and expensive to create 360° views of customer engagements and the business.
4. Implement rigorous authentication and access protocols. If not designed correctly more people will have access to more data which can create exposure from multiple different fronts around confidentiality, privacy, business sensitivity, etc. Your corporate security policies need to be updated and implemented to reflect these new realities.
5. Build a seamless user and data experience across data analytics and IT platforms. If your customers, even your internal staff have to struggle between multiple systems, adoption and growth of your IoT business will be limited. It is important to make it easy to do business with.

If integrated right, the growth pace of your IoT business will get accelerated and the cost to support that growth will come down.

9.8 Engaging Your Stakeholders Through Reporting

There are a number of different stakeholders that are involved with your IoT business:

- Customers
- Shareholders
- Other businesses whose products are being IoT-enabled
- Channel partners
- Your IoT business team

As you build your business, it is important to keep them engaged and informed. You need well thought out reporting cadence to help with this.

Reporting helps in many ways:

(a) Increases top of mind recall
(b) Improve confidence that your customer and install base is growing
(c) Demonstrate value creation
(d) Establish your significance for your channel partners as the right gateway to customers
(e) Invite other ideas and investments for growth

Different stakeholders care for different information and metrics. You should have periodic standard reports which address those needs and have capabilities to answer ad-hoc reporting requirements. You need to exploit every opportunity to highlight the metrics that exhibit your growth and share it extensively across multiple stakeholder groups. These include:

- Install base
- Customer base
- Number of offerings
- Wallet share
- Value delivered for customers
- Volume of data
- Analytical capabilities
- Intellectual property
- Channel access
- Revenue and income

All of these should be shown as growth over time. It will be helpful to also show events and interventions along that time scale to provide context to the growth story.

9.9 Summary

In this chapter, we explored a few elements of how to close the loop with customers to deliver value using your IoT products and solutions. In previous chapters we talked about the key delivery components; in this chapter, we delved into the business support ecosystem which facilitates actualization of the delivery components. This typically includes capabilities for customer management, managing partners who are contributors to your IoT ecosystem, managing commercial transactions, business reporting, and financial reporting.

As your IoT business continues to grow, the operational expenses will start to pile up. As per a Cisco study, nearly a third of the expenses could go towards network communication, administrative activities, and technical support. These depend on the amount of data transfer, the extent of analytics and technical support provided. Some of these will come down as scale goes up, communication and data storage costs come down. The business support ecosystem needs to accommodate these aspects of your business. It is important that the ecosystem supports you to launch new services quickly and streamline operating costs.

You can outsource many of the business support activities, but you need to have substantial control to influence your success through better customer engagement.

Often companies invest a lot of money and thought in building the technology infrastructure of the IoT ecosystem. However, business benefits including appropriate monetization can only be enabled by the business support aspects of the ecosystem.

References

1. (Ma, et al., 2014) Ada Ma, Youngchoon Park, Sudhi Sinah, Jignesh M. Patel, Self-Conscious Buildings: Transforming Buildings in the Age of the IoT, Johnson Controls Business Strategy Report, Jan, 2014
2. (Daryl, 2003) Daryl D. Green, 2003, retrieved from https://nuleadership.com/2013/10/14/customer-motivation/
3. (Carmine, 2015) Carmine Gallo, 2015, retrieved from https://www.forbes.com/sites/carminegallo/2015/04/10/how-the-apple-store-creates-irresistible-customer-experiences/#29875c6417a8, Forbes.com

Chapter 10
Dealing with Security, Privacy, Access Control, and Compliance

10.1 Introduction

IoT has dramatically enlarged the playing field with devices and data. While this brings many benefits, it also creates significant exposure to security and privacy vulnerabilities. IoT amplifies the access points for data and control, which in turn amplifies the intrusion points [1–3]. While we continue to build defenses in devices and networks, we also have to deal with a huge population of legacy devices, applications, and networks where inbuilt protection was limited [4, 5]. The threats are becoming more persistent and the impact more profound, sometimes debilitating to businesses. The threats are equally high for consumer IoT businesses as well as industrial IoT businesses. History is riddled with many examples of security breaches with significant impact. The discovery of Stuxnet [6] in 2010, a small 500 kb worm that infected the software for 14 Iranian nuclear power plants brought a lot of focus to this subject. However, there are examples from before. The Slammer worm disabled the Davis-Besse nuclear power plant in 2003. We continue to hear stories about credit card and another personal information breach all the time. A 2014 SANS survey reported 7% more respondents indicate a breach of their environments [1].

The impact of these security breaches is manifold, you experience:

1. Economic impact arising out of business disruption leading from non-functioning goods and services
2. Public safety concerns
3. Reputation loss and brand damage for companies
4. Increased investment in product development and maintenance
5. Intellectual rights get breached

This is a complicated subject because there are emotive, ethical, legislative and commercial considerations in addition to the technical complexities. On one hand, businesses do not want to compromise on security, on the other hand, they do not

S.R. Sinha, Y. Park, *Building an Effective IoT Ecosystem for Your Business*,
DOI 10.1007/978-3-319-57391-5_10

want to be slowed down by this rapidly evolving space. In this chapter, we will broaden your horizon about security and privacy, and give you practical guidance on how to build barriers against threats.

We shall cover the following topics in this chapter:

1. Understanding the threat vectors in the world of IoT
2. Recognizing data privacy and ownership concerns
3. Learning about the various industry standards and available resources
4. Understanding popular and modern technologies and standards
5. Designing security policies and practices
6. Securing data and insights
7. Building risk management and compliance program
8. Hacking and auditing security and privacy effectiveness'
9. Managing attacks and communicating effectively

Before we get deeper into the topic, let us introduce you to three terms we will be using in rest of the chapter:

1. Threat vectors—These are the paths taken to penetrate your IoT devices and software programs with intent for malicious content delivery. Treat vectors includes tools and methods used to accomplish the breach.
2. Threat actors—These are the perpetrators who engage in security and privacy breaches. They are often humans and sometimes software programs.
3. Threat surface—This is the extent of exposure in your product, services and install base to specific threat vectors. This is very dynamic; you are engaged in a constant battle to reduce threat surface while threat actors are working on increasing it.
4. Malware payload—These are software programs that cause damage to your computational activities, corrupt data or steal information.

10.2 Understanding the Threat Vectors in the World of IoT

Unlike traditional cyber security threat management technology where these issues revolve around software and how it is implemented, IoT concerns what happens when the cyber and the physical systems converge. Protecting IoT solution requires a holistic view of threat management around the connected device, communications, cloud services and users (Fig. 10.1).

There are different types of threat vectors, keep in mind, more gets added with time as new vulnerabilities get created.

Some of the current popular ones are [5, 7, 8]:

1. Communicating devices—Many in the rapidly growing population of communicating devices that become part of your computation or processing network, do not have the appropriate high levels of security implemented on them. This becomes an easy path for intrusions.

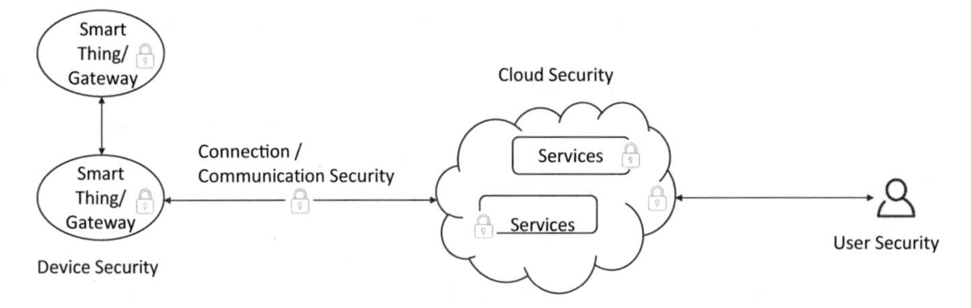

Fig. 10.1 IoT Security Solution must address holistic view of threats

2. Removable media—Removable media like USB devices are often used to download content to your computing device. This is an easy route to deposit a payload of malware into your device and subsequently transmit it through the network.
3. Networks—Once a payload of malware enters a network somehow it can travel across to create damage. While unsecured networks are more susceptible, even secured ones are not safe as new dormant malware gets introduced. Lack of appropriate firewall control of IoT networks, including lack of sufficient logical separation from enterprise IT/OT networks or internet further accentuates the problem.
4. Email inclusions like links and attachments—Malware gets downloaded into the device or network through these links or attachments. By default, we trust most of the content we receive unless there is an obvious issue. The issue becomes severe because often attackers hack into the email system of our known trusted network and transmit malware through their emails without their knowledge.
5. Websites—These are becoming very common in the form of ads, banners, cookies, and bugs which we inadvertently engage with.
6. Identify sinks—There is a lot of identity and personal information stored in social media, online financial transaction systems, and similar applications. Once breached they become an easy route to access the network of other people and devices for transmitting malware. An example is Spoofing identity that someone or some device using another person's or device's identity.
7. Tempering with Data—Altering the data related to a device or traversing the network.
8. Remote access—Weak security controls for remote access including internet facing systems is another growing threat vector. Vendors and contractors, VPN configurations, the use of personal devices and vulnerable operating systems, where you can apply limited control become another popular route for hackers.
9. Access rights—Provisioning users with elevated privileges beyond the minimum required, such as the use of administrator accounts for routing functions,

creates a risk for unintentional and malicious incidents. An example is Elevation of Privilege that an unprivileged user or a device gains privileged access and thereby has sufficient access to compromise or destroy the entire system.
10. Intelligent viruses—These malware can spawn more of their kind, not only limited by the initial payload of malicious content.
11. Code vulnerabilities—Sometimes due to weak programming practices like a stack overflow or other bad coding methods, we leave vulnerabilities in the code itself. Threat actors leverage these to introduce malware payload into the system.
12. Denial of Services—Disable a particular service via requesting too many resources.

Along with securing a device and connectivity that we discussed in previous chapters via device management and various protocols, we must secure an ecosystem that requires a rigorous strategy. We must be able to ensure and validate clean code loading into a device from the manufacturing process. An integration partner or a solution developer must practice design for security based on available standards.

A deployer of IoT solution must have traceable evidence on hardware settings and commission information of networking, user access modeling, and other security sensitive policy. Lastly, solution operator shall perform regular monitoring and analysis of ecosystem's security health.

10.3 Learning About the Various Industry Standards and Available Resources

For complex topics as cybersecurity, it is important to leverage frameworks to pivot our thinking and actions. There are many frameworks which can help you work through cybersecurity issues. The NIST Cybersecurity Framework [9] is a good one to start with. It is quite exhaustive and comprehensive. This will help you build a context around your scope, exposure, and actions. You can find more information about it in: https://www.nist.gov/sites/default/files/documents/cyberframework/cybersecurity-framework-021214.pdf

Early work started on cyber security in 1998 at the Stanford Consortium for Research on Information Security and Policy in collaboration with US National Security Agency [10]. Around the same time, UK Government's Department of Trade and Industry published the BS 7799 [11], part 2 of which dealt specifically with information security management systems [11]. Initially, the focus for many such efforts was focused on critical infrastructure. In the last 10 years, the threat exposures and attacks have become more prevalent and popular. Consequently, there has been a lot of collective thinking around standards between industries, governments, and academics. The table below captures some of the popular standards. We have not gone into details of explaining each standard and how to implement them here in this book, you can such information easily on the websites of sponsoring organizations (see Table 10.1).

Table 10.1 List of popular security standards and guidelines

Standard (version as of 14 Jan 2017)	Sponsoring organization	Key highlights
ISO 27001 and 27002: 2013 [12]	International Standards Organization	This is the most popular and widely accepted set of standards for information and cyber security. These are very comprehensive and captures best practices across every dimension of security that impacts your business. 27001 covers the 'what' and 27002 covers the 'how'. 27001 is auditable and 27002 is the enabler. These evolved from the BS 7799 standards. Many companies use these to demonstrate their thoroughness in the adoption of security practices.
TC CYBER 2014	ETSI Cybersecurity Technical Committee	Designed by the European Telecommunication Standards Institute, these are globally accepted but nearly mandated by most European nations for businesses based out of or operating in Europe. These focus heavily on communication and privacy aspects of IoT businesses.
Standard of Good Practice	Information Security Forum	This is an extremely good standard for businesses to adopt. While it closely aligns with the ISO standards, it also provides practical guidance on managing cybersecurity programs. It also combines the needs of other compliance requirements like COBIT, Sarbanes-Oxley, Payment Card Industry Data Security Standards amongst others. Following these practices help you get better coverage across the spectrum.
ISO 15408 3.1 rev 4 [13]	International Standards Organization	Integration and testing of multiple software applications. Component developers and vendors can comply their offerings to provide assurances to other applications/customers using them.
ISA/IEC 62443	International Electrotechnical Commission, International Society for Automation	Comprehensive standards and guidelines for control systems. These are broadly applicable for most industrial internet applications. This covers component suppliers, system integrators, and end customers environments.
	American National Standards Institute	There is a lot of good details around policies and procedures that relate to various aspects of development, installation, and maintenance.
NERC CIP 002 through 009 [14]	North American Electrical Reliability Corporation	Extensive guidelines on how to best protect networks for critical infrastructure supporting systems. While it started out for large scale electric infrastructures and networks, the concepts are extensible to other similar networks.
NIST-800	National Institute of Standards and Technology	These are used widely by the US and other government agencies. They are fairly comprehensive and also include elements or privacy and social concerns. Any business wanting to work with the government must take these seriously. These also include guidelines on how to safeguard personally identifiable information (PII) which is today mandated by law in many jurisdictions.

While there are many common themes across all of them, there are some unique dimensions to each. It can be very confusing which one to follow and use as the basis of cybersecurity program for your IoT business. You should weigh your choices against the following parameters:

1. Applicability to your business
2. Acceptability in the markets you serve
3. Requirements of the jurisdictions you operate in
4. Competitive action around standards adoption

A good rule of thumb is to start with a review of the ISO, NIST, and Standard of Good Practices. Many private companies which operate in the IoT space like Microsoft and Cisco have evolved good standards and practices as well. If you have any engagement with them, it will be a good idea to do some knowledge sharing and benchmarking.

This is a journey. It is important to start somewhere and adapt as your business evolves. It is important that you familiarize yourself with the various options and keep up to date on the improvements to your chosen ones and look out for new ones.

10.4 Recognizing Data Privacy and Ownership Concerns

IoT is really a combination of products, technology, data, analytics, services, and outcomes. This makes the data privacy, ownership and access considerations very complex. Moreover, there are different implications and concerns for different industry segments like healthcare, consumer, industrial, smart communities, public service delivery, etc.

There are several dimensions to this topic.

10.4.1 Data Ownership

IoT has transformed data into new currency and source of major economic value creation. Data can be valuable for direct analytical purposes, and it can be sold or traded for value. Companies engaged in IoT space are incorporating contractual clauses to establish data ownership. Ownership of data is important because appropriate license rights may provide an entity with everything it needs to conduct its analysis or monetization activity. Each entity will need to determine how it approaches this own versus licensed data. Some IoT businesses will need to develop playbooks for scenarios where major clients are unwilling to allow the IoT solution provider to own the data outright. One option is for the IoT company to have a perpetual license to the data so long as any personally identifiable information is removed from the data during further processing steps.

10.4.2 Rights Around Derivative Use of Data

Often, collected data can be used for purposes other than its original intent, without express consent from the generator/owner of the data. For example, the users of health/personal IoT devices like Fitbit may have allowed the equipment/service provider to take data for giving a summary of lifestyle behaviors. Now an insurance company can acquire access to this data and modify the health insurance premiums based on the lifestyle behaviors—the device wearers may not have consented to this. Such a scenario could cause IoT adoption to be a disincentive to the user and pose regulatory hurdles around discrimination and denial of normative services. However, anecdotal evidence is that consumers are becoming more used to exposing data in exchange for functionality or analytical value in products and software.

10.4.3 Dynamic Decision Rights

Typically, when people and businesses first start using IoT devices, the manufacturer or service provider ask for their consent to use data generated from their devices, and possibly even to share it with others. Even though people provide this consent at the onset, they might have second thoughts later due to evolving comprehension and preferences of how their data might impact their well-being or a change in perception about the value they expected from the device or service. Hence, the consumer might want to retain the right to un-consent and be forgotten. We will see more recurrences of such instances where IoT interfaces with people's health and financials data.

10.4.4 Consumer Awareness

Often, users of IoT devices, as well as service providers, may not completely understand the nuances of technology and data analytics. Considerations mentioned above are examples of that. Today, manufacturers or service providers aren't required to educate consumers about these implications or set up advisories, which is necessary for the space of healthcare, insurance, banking and several other industries. IoT is quite pervasive these days, its impact and implications are far reaching. So there is a need for better practices around providing transparency to consumers.

10.4.5 Privacy Rights

Starting with large search companies and social media, and now with IoT, we are always connected—and our every action is traceable. Privacy and anonymity are severely impaired in this environment. This could lead to infringement of personal freedom and even expose people to higher risks of fraud because their personal data is easily available. There is increased activity around personal information, in many cases covered by law of the land, something that is complex to consider given the expanse of reach of IoT devices and applications.

10.4.6 De-aggregation Standards for Data Privacy

A common way to improve privacy is to de-aggregate data and anonymize it. De-aggregation breaks up a logical group of data into isolated pieces which in itself may not make much sense or help create any value. For example, in an online purchase—capture, transport and storage of user's name, address and credit card information happening separately is de-aggregation. If the transport and storage of data follows different routes, it creates a barrier for any hacker and prevents any unintended use of data.

10.4.7 Liability

Data, analysis and control actions change multiple hands in a typical IoT ecosystem. For example, a wearable device might capture heartbeat and few other indicators, then transmit it to some processor of the data, which uses algorithms from another provider to calculate health risks and need for emergency response, and sends it to a healthcare provider, who in turn uses a different service for patient monitoring. So, if something goes wrong, accountability around outcome becomes a very tricky situation, because it is hard to determine what really caused the failure. Even if it is clear, the different participants might have different thresholds for accepting a share of liability as their levels of economic benefit in the IoT processing chain is different. The legal framework from the industrial age handles single accountability scenarios better, but we are fast moving away from that world.

10.4.8 Reliability and Accuracy Standards

Already a number of critical services like healthcare monitoring and weather monitoring are transitioning to more IoT-type devices. This impacts people's lives and livelihood. So there are expectations of an extremely high level of reliability and accuracy of data coming from IoT devices, as the central processing algorithms and control functions rely entirely on the incoming data. For example, due to some communication break, if a pacemaker fails to transmit data for a certain period, it should not turn off; if a drug delivery system receives wrong readings from sensors attached to a human body, the effect could be fatal; if sensors collecting pollen count are not calibrated right, it can lead to incorrect allergy reporting, which can result in serious health issues. To make IoT devices for all such applications affordable to increase their infiltration, we run the risk of compromising on their reliability and accuracy of data capture, data transport and local compute capabilities. We need standards on how to balance affordability and quality.

10.4.9 Trustworthiness

IoT applications are not limited to a single device or service, it is usually part of a larger and wider communication fabric where data is generated, captured, transported, analyzed, processed, insights derived, actions were taken, the value captured and reported. For such an intricate system to work with multiple hand-offs, it is critical that safety, security, and resilience of data between the different participants and processing chains are maintained. If the data, insights, and actions cannot be trusted without a doubt, and each step of the processing chain or participant has to account for gaps, the benefits and value will get compromised. This is similar to the banking and financial industry in many ways. Only accepted industry standards or government recommended policies can help address such trust issues.

You will have to think about building responses to these issues in your IoT business design.

10.5 Understanding Popular and Modern Technologies and Standards

There are several ways to protect data and computation. Here we explore some of the popular ones employed by different businesses.

10.5.1 Encryption

Encryption techniques evolved from cryptography as the method for encoding messages or data content in a way as to avoid unauthorized access. Encryption always requires a key to encode and decode a message. While the basic tenets of cryptography date back thousands of years to Greek civilizations, modern encryption techniques date back to the early 1970s and have evolved in sophistication since then. Commonly available encryption methods and tools have been available in the computing world since the 1990s with the rise of malware and intrusions. With more data getting exchanged between devices and networks in IoT, encryption is on the rise for data in transport. Advanced Encryption Standard (AES) published by the

US National Institute of Standards and Technology (NIST) in 2001 started convergence around generally accepted standards in the industry [2, 15–17].

There are two primary ways of encrypting—using public or private keys. The primary difference is in the construction of the deciphering keys. When same or symmetric keys are used, it is private; when disaggregated asymmetric keys are used widely, it is public. Complex mathematical algorithms are applied in both cases. Encryption levels are made more stringent by using higher bit size keys for encryption and decryption. 128-bit encryption is quite popular in most cases, but for more critical applications, we are starting to see higher numbers like 256 or more. If you are using a 128-bit encryption, to break that you need to try 2^{128} combinations, which requires considerable time and computational power. However with increasing processing power and reducing costs, breaking these higher encryption levels is becoming possible. So other methods like blockchain are being explored.

RSA algorithm [18] developed by MIT scholars Ron Rivest, Adi Shamir, and Leonard Adleman is one of the popular public key encryption techniques. It splits the entire process into four steps of key generation, distribution, encryption, and decryption involving multipart algorithms. Blowfish and Twofish are popular private key encryption techniques along with AES [18, 19].

There are many challenges and disadvantages of encryption techniques:

1. They are inherently complex and required specialized skills to implement
2. They create computational and administrative overheads, an issue for IoT applications given the high density of computational points and high throughput required
3. Given enough time and computational capabilities, they are vulnerable to being deciphered without exactly knowing the decryption keys
4. If not implemented well and kept current with new threats and methods, you will be lured into complacency
5. They are less effective in cases of large population of distributed devices where key management can be cumbersome and susceptible to other threats

10.5.2 Malware Analysis

Malwares are software programs which either intent to harm the functioning of its host or steal data from its host. There are many types of malware—spyware, virus, worm, trojan horse, ransomware, adware, rootkit, or backdoors. All of them are usually self-contained, generally invisible and appear innocuously. They could show up as part of an ad campaign, an email attachment which you downloaded, a browser cookie, or any similar normal action. Trojan Horses are actually part of a legitimate event, the malicious content getting activated by a later trigger. Some of them like viruses and worms have self-replicating capabilities making them hard to contain. Malware like rootkit has a disastrous impact as they modify the underlying OS layer. In many cases, the malware will not show any symptoms for a considerable period of time after infecting.

Malware first appeared 30 years back. Viruses were the first to appear in 1986 with the name Brain, developed by Pakistani brothers Basit and Amjad, soon followed by the first worm accidently developed by Robert Morris in 1988 [20]. Stuxnet which is referred to earlier in this chapter was the first of a new class called super malware, it used multiple forms of malware working in coordination, had intelligence, self-generating authentication, and self-destruction. Today Malware is a large enterprise and costs billions of dollars in economic impact.

There are many techniques to identify, isolate and endorse against malware. Anti-virus, network sniffers, debuggers, decompilers, and disassemblers are often employed for malware defense. For any known malware, you need to have a standard incident response mechanism. Every time you encounter a new one, update the same plan. When you see a new malware:

1. Understand the threat vector or path taken by it to cause the infection
2. Identify the threat surface
3. Figure out how it works

This will help you build a proper response method.

You need to understand the malware both at rest and in action, so you need to do a code analysis to the extent possible, as well as see its functioning in a contained test environment. Sometimes it can be very difficult to replicate a malware's functioning and understand its code because of inherent complexity, like in the case of Flame which had organic growth capabilities. Every analysis on a malware must be performed in a clean-room type environment where there is no connectivity to any sort of outside network, device or application, sometimes benign malware will infect if the malware management system if not properly isolated. Today there are a lot of tools available for malware detection and handling. You will find a lot of open source solutions for this, but it is prudent to validate their authenticity and effectiveness before you choose any of them. Use your in-house cyber security team, IT department or the services of a reputed security consultant to help you make that determination.

10.5.3 Intrusion Detection Systems

Intrusion detection systems (IDS) are independent software programs or embedded in a special purpose device that monitors and respond to security breaches and malware. Most networks these days have some form of IDS. While firewalls look out for any threat and prevent their entry into your network, they are limited by a range of known scenarios. IDS can adapt and investigate situations which have sneaked into your network. IDS can be applied at a network, individual device or computational block level. IDS looks for signature patterns of known malware or applies machine learning techniques for unknown anomalies.

IDS may have limited value in certain scenarios like:

1. Understanding false alarms created by bad code or data
2. Encrypted malware contents
3. Time delayed malware or outdated signatures where no patterns are available
4. Targeted attacks
5. High computational resource needs as the threat library grows

10.5.4 Blockchain

The security specifications around IoT systems involve multiple topics including, securing messages sent on the communication media, user/operator authentication and authorization, entity identity and integrity of the data. Typically, authentication and authorization in current IoT devices and systems are implemented in a centralized manner. Central security server based policies become a single point of failure in the design and make it vulnerable to attacks and faults in the system. IP or wireless-enabled IoT networks and cloud technologies bring more reliability and scalability; however, the same flaw for centralized security design still holds. The data can be manipulated or the identity of the entities can be forged. The communication overhead with the server and any encryption-related processing is another disadvantage that results in efficiency degradation and extra maintenance cost.

Blockchain technology address many of these issues. It leverages a decentralized and distributed database of continuously growing order encrypted records called blocks which are commonly shared and maintained across the computing network. The peer-to-peer and distributed timestamping allows for secure autonomous management of the blockchain database. The data and communication security uncertainty are significantly reduced by blockchain technology due to mass collaboration based authentication. You can use blockchain for:

1. Transmittal of access rights and licensing information
2. Transmittal of actions that can have significant impact on the IoT system
3. Transaction negotiation between different machines
4. Data transport of sensitive information across the computing network
5. Any communication in highly sensitive industries and applications

There are several advantages to using blockchain technology in securing. To name a few:

1. Significantly higher levels of security as same information is held across different nodes in the network making it nearly impossible for any threat vulnerabilities as you have to simultaneously penetrate all nodes instead of attacking just a central server or failure point
2. High byzantine fault tolerance due to distributed nature
3. Mass disintermediation due to consensus-driven decision-making

4. Resistant to data modification due to public sharing of block information and tokens; no retrospective alteration of data
5. Allows contractual transaction at machine level without any human intervention
6. Enables auto generation of high secure communication and transactions in an agent enabled intelligent network

The blockchain is relatively new was initially used by the Bitcoin movement and later applied to many financial applications. While using blocks in cryptography started over 25 years back, the present concept of blockchain by Satoshi Nakamoto dates back to 2008 [21]. We believe IoT and blockchain technology will experience synergistic growth.

10.5.5 Context-aware Behavior Analysis

This method uses pattern driven machine learning to understand potential threats and malware. It does so by tracking behavior and functioning of users, devices and communication networks. You can also analyze software and file structures to predict potential vulnerabilities. This can be very effective as you do not have to wait till the event happens but can identify the leading indicators. Sophisticated threat actors will typically engage in early activities to understand the vulnerabilities and breach smoothly; such instances are more than full-force attacks.

Some of the leading indicators which you can use for modeling and detection are:

1. Location of generation of authorization or access request. This could be geographic locations, browser types, unrecognized devices, command line or system level access attempts outside of system administrators and such other factors.
2. Unnatural system access and usage by otherwise authorized users. This usually indicates somebody else has breached a user's credentials and is using them with malicious intent.
3. Traffic hopping. If you are directed through multiple different paths to your destination or if there are too many new unrecognized hoops in the network traffic, that is an indicator of somebody diverting communication traffic with an intent to breach and cause harm.
4. Unnatural device behavior. If an IoT device is detected to be working in a manner which is not normal, say your Fitbit is showing erroneous readings despite being fine otherwise, or a smart thermostat is showing incorrect temperatures but otherwise seems to be working fine, there is a high possibility somebody has breached it and playing with it to damage something.
5. Illogical system growth. Sometimes you can see files and programs growing in size or content without any particular reason. This could be because there is a hidden malicious payload hidden and progressively constructing itself. Slowing growing.

The possibilities are many and you have to keep learning and defending. Often the paths of analyses are not linear, you may have to explore a combinatorial effect. You also need to analyze in layers of your security implementation; Avivah Litan has created a 7-layer model for the same. This approach is also helpful to weed out the false alarms generated and draw your attention to the most important ones that need some subjective judgment and possible human intervention. However there is one limitation with this approach, you need a lot of data to create meaningful insights.

10.6 Implementing Security Policies and Practices

Given the far-reaching implications of security in IoT, all the stakeholders in an IoT business—manufacturers, offering developers, service providers, and customers have to participate actively in the implementation of security. It is critical to apply appropriate measures right from business and product design phase. Security at the design phase has to be incorporated in device access, user access authentication, network topology, upgrades and service events and finally obsolescence. Obsolescence is often ignored but is very important in this case because technology is constantly changing, with increasing demands of functionality devices will run out of capacity at some point, and new threats and vectors will keep emerging. So you need to take a full lifecycle view of security.

During the design phase, there are a number of steps you need to follow to account for security considerations. Here are some of the key imperatives:

1. Assess threat vectors and vulnerabilities. Map out the intended use of your IoT device and system, potential unintended usage and consequences, coverage of possible actions, connectivity options, and implications, and what makes the devices and applications susceptible to threats. You need to do the assessment at an individual device/application level as well as the system level to obtain a comprehensive view. It is a good practice to assign threat levels and the likelihood of occurrence while you are doing these assessment to understand priorities.
2. Designing protection strategies. For each identified vulnerability, you need to next design how to defend. If there are multiple threat scenarios for a single device or application, you may need multiple defense postures.
3. Simulating breach responses. Once you have identified the vulnerabilities and possible defense approaches, you should simulate the sufficiency of your protection strategies. For example, you may have a device which works in a wireless network and your defense relies on firewalls and communication encryptions to and from that device. If you have a USB connection to that device but do not have any IDS implementation, your security design is flawed.
4. Creating secure development policy. This is a very important step to bring institutionalized disciple across your development organization around security.

The policy needs to establish the principles and authorities for implementation of security practices in developing customer-oriented offerings. It should include purpose, scope, various roles and responsibilities, information classification schema, training coverage and methods, design processes, and review mechanism. The policy should be broadly reviewed and supported by the business and technical leadership of the organization. To be effective, the policy has to be inviolable with clearly laid out consequences.

5. Developing secure development practices. Based on the policy, you need to now develop specific standards and guidelines for your offering development. This will include both hardware and software based products. The secure development practices should address how to secure communication ports in different types of IoT devices, connectivity best practices, how to safeguard various network ports and intrusion points, network segregation, firewall rules, cloud connectivity, user and content authorization protocol, password management, remote device and application access, router access, remote control capabilities, firmware protection, protocol conversion, data transmission activities, encryption methods, operating system level considerations, runtime environment handling, software and firmware upgrades, patch management, programming standards to leave no vulnerability in the code itself, third party software and device integration, managing audit logs, and malware protection implementation. Collaborative effort www.safecode.org has a lot of good resources to help you develop specific development practices.
6. Training on secure development policy and practices. For robust implementation, it is of paramount importance to training your entire staff involved with product development and support to learn about the secure development policy and practices. This will help raise awareness and capability levels of your organization around security implementations. You will most likely have a set of web-based training on the policies and practices. You should augment those with interactive sessions where the developers and other stakeholders get an opportunity to clarify queries and concerns. Training should be mandated and refresher courses a required criteria for continued engagement with product development and support.
7. Implementing secure development policy and practices. Once training is underway, for effective deployment, you need to integrate your secure development policy and practices into your normal product development process. Add the relevant checks and balances during requirements, design and validation phases. Similar routines should be implemented for products and services under maintenance where little to negligible new development is occurring.
8. Testing for existing and new vulnerabilities. Next, you need to test the effectiveness of your practices against the various vulnerabilities through black-box and white-box testing. Threats, vectors, and actors keep evolving with time, so you need to keep up with them. Look out for new vulnerabilities that are coming up. A good place to start identifying them are the voluntary disclosure platforms and government cybersecurity advisories. Once you identify them,

design a remediation or protection mechanism and test those in a simulated environment.

9. Revising threat vectors and vulnerabilities inventory. As you learn about new vulnerabilities and better ways of dealing with them, you need to update your knowledge base with these inputs.
10. Updating secure development policy and practices. After updating the inventory, next update the policy and practices. This will help your implementation be current and IoT business is safer.

As you have seen so far, managing security and privacy is a complex endeavor in the IoT space. Moreover, it is ever evolving as is your IoT business. So it is recommended that you make a visual representation of your IoT business in terms of the product lifecycle. Then represent the various threat vectors and threat scenarios with possible implications in that lifecycle model.

Similarly, take another view of this with the user as the pivot. Map the various customer engagement interactions users have with your products and business and show the product security interfaces in that map.

10.7 Securing Data and Insights

Often a lot of effort and thought goes into securing the access to the system and better programming practices to eliminate vulnerabilities. These are usually adequate in ensuring security for enterprise type systems. With IoT, you also have a deal with a lot of personal information dispersed across millions of users, devices, applications and locations. Personally identifiable information (PII) has become a very hot topic in the IoT space. PII includes any information that can identify a unique individual. Such information can be maliciously collected through breaches into devices, browsers, and networks. Breach of such personal information is a huge concern for users, especially with the rising incidents of data theft. Nearly all IoT business, consumer IoT more than industrial IoT, collect PII. So securing such information becomes a top priority.

On similar lines, many corporations and businesses are also very sensitive to some business data and insights generated from data analytics. These could be client confidential and competitively vulnerable. However, they may need to interact with external systems and suppliers or customers with data in the context of their IoT business. You may also be limited by contractual obligations of not sharing data without encryption or anonymizing when going outside of your company. When using 3rd party platforms for your analytical processing, you need extra care because you have no control over your data in other environments. Data and insight security in such instances become very critical as well.

There are many ways to secure data:

1. Fully encrypt and/or anonymize the data at the point of generation and have non-identifiable data come to the analytics platform.

2. Encrypt personal or sensitive data at the point of generation for safety and use partial strings for further storage and processing.
3. Use encryption at the data generation point has the keys secured dynamically within the data storage and analytical platform so that there is no exposure. Right after processing, re-encrypt the output data and insights. This way, it will be near impossible for humans or software programs to breach.
4. Secure after you receive the data provided you have confidence in the security of the data transport mechanism.

You can put layers and layers of encryption and anonymity at different stages of data transport and processing. You need to evaluate your approach based on business implications and processing hassles.

The challenge with normal encryption and decryption is that there is usually a window, however small, where you leave your data vulnerable to attack. One way to address this is homomorphic encryption. This is a method for computational activities without decryption, so imagine that your data which does not make sense to anybody else right now is taken through an analysis and processing chain with only the processing chain understanding the content of the data string. You can do partial or fully homomorphic encryption, the more you do, there are computational complexity and overhead. For partial homomorphic encryption [19], there are several popular methods developed by experts like Tahel ElGamal, Shafi Goldwasser and Silvio Micali, Josh Benaloh, Pascal Paillier, Tatsuaki Okamoto and Shigenori Uchiyama and many others. For fully homomorphic encryption, Craig Gentry is usually accepted as the leading expert. Except for the Gentry method which is less than a decade old, most of the other methods are older and have been applied in several business applications. IoT opportunities have magnified their implementations.

There is a lot of advancements being made in crypto research and we will continue to see better tools and methods.

10.8 Building Risk Management and Compliance Program

Companies have been building financial risk management and disaster recovery programs for several decades. On similar lines, you need to build a risk management and compliance program for cybersecurity to support the growth of your IoT business. This is a very involved process which should be thoroughly documented, communicated, implemented and audited. You must have an executive engagement to outline organizational risks, set priorities and empower decision-making process. It is better to develop policies, processes and perform resource allocation within the operating business units as they are closest the market, customer, and technology realities.

To begin with, you should constitute the following teams for managing cyber threats, attacks, and risks to your business:

1. Security Governance Board—composition, charter, authority
2. Security Council
3. Product Security Champions
4. Incident Response Team

Upon setting up these teams, following are some recommended steps to build out your risk management and compliance program:

1. You must designate somebody in your organization to be in charge of cybersecurity. In some places, they are called Chief Information Security Officer or Chief Product Security Officer or other similar names. This person should be empowered with policy formulation, implementation, and compliance reporting. This pivot person needs to work with the previously identified groups and functions for daily operations and periodic reviews.
2. Next, you must define your compliance framework. This should include your secure development policy, privacy rights, access and authorization control mechanisms, regulatory and compliance requirements required by the law of the markets served. Often data residency is a key aspect of such compliance in European nations and countries like China, especially for data which contains personal information and may be construed to contain information of national value.
3. Train people on compliance framework. Even if individual job scope covers only limited aspects of the framework, still give everybody exposure to the entire framework. This broadening will expand the limits of your defenses.
4. Integrate compliance program into your product/service development and maintenance programs. This way you can avoid additional overheads and delays in launching your products and offerings into different markets.
5. Design and implement an audit program. Best audit programs are a combination of standard certification programs and a number of self-assessment initiatives. We shall go more into the audit aspects in the next section below.
6. Finally, outline the reporting cadence around security compliance. This should include the content and schedule for internal and external reporting. Voluntary declarations are becoming more and more popular. If you make that choice, you will need to internally brief the appropriate stakeholders prior to any external filing.

Another key aspect of your risk management program is educating your customers and stakeholders on the issues around cybersecurity and privacy. However good your practices and implementation might be, its efficacy will be limited by the weakest link in your customers' environment—be it an individual or a large corporation. If somebody shares their password or allows any malware, your burden goes up significantly. So you need to educate your customers like you do for to your staff, on the perils, implications, and defenses around cyber security. Likewise, you need to educate your shareholders and other business stakeholders about these topics so that they can understand what and why if you ever have an incident.

10.9 Auditing Security and Privacy Effectiveness

Reviewing the effectiveness of your security and privacy assurance program is as important as designing and implementing them. There are many ways of doing that, you can even employ multiple methods simultaneously.

1. Measuring training effectiveness—Training becomes the basis of your implementation, so knowing its effectiveness is a key step for the validation exercise. Training effectiveness should be triangulated between direct training feedback from participants, listing avoidable exposures and successful defense. Any one of the techniques will give you biased results.
2. Vulnerability testing—This is testing against known threat vectors, actors and threat surfaces against your policy, practices and other established industry standards. You should have some level of internal capabilities and tools for such testing but should leverage specialized external capabilities to get a more objective and thorough view.
3. Ethical hacking—Today ethical hacking is considered as a legitimate and important step in security compliance programs. Many large companies have in-house ethical hacking teams like IBM X-Force Ethical Hacking team [22]. If you do not have the scale, you can retain the services of individuals or organizations that are certified by the International Council of Electronic Commerce Consultants popularly known as the EC-Council. EC-Council administers the Certified Ethical Hacker test.
4. Honeypots—These are decoy devices, servers or networks used to lure intruders and threat actors based on perceived vulnerabilities. Honeypots serve multiple purposes—you can identify individuals or organizations that are a threat, you get first hand and a better understanding of intrusion events, you can run simulations, you can test your protection tools and techniques. However, these might be legality concerns around deploying honeypots in certain jurisdictions of events because you are essentially inviting somebody to commit an offense.
5. Standards conformance certification—These are universally accepted endorsements for your security policies and practices. Depending on the nature of your industry or markets served or a combination of both, there are several options available. For example, IEC 62443 Conformance Certification is widely accepted as a gold standard for industrial automation. Likewise, UL Cybersecurity Assurance Program is becoming a popular model for multiple areas spanning industrial controls, home and commercial office automation, consumer goods companies amongst others. ISO 27001/2 and NIST standards also have audit options [10–13].

It is good to have some audit program, not only does it help you with PR, but also it helps with internal confidence building and learning.

10.10 Managing Attacks and Communicating Effectively

As much as we would like to avoid, but if operate in the IoT space long enough, the chances of being the victim of a cyber attack are always there. So far we have discussed extensively how to prepare defenses against such attacks. Now let us discuss how to manage incidents. There are a few steps you should follow for every cyber attack:

1. Activate your incident response, team. Intimate them about the event with as many details as you can provide. All members of the team should know their specific roles and actions.
2. Understand the attack surface, threat actor, and potentially leveraged threat vectors. This will give you an indication of the gravity and potential impact of the attack.
3. Explore your response options. Investigate whether you have a response plan for the given situation and then activate it. Sometimes you may be limited in exercising the response action by the attack itself or potential negative implications to your customers and business. In such an event, explore other ways to safeguard your customers and your interests.
4. Communicate internally and to customers. Both are equally important. Internal people will help with response, reporting, and future risk avoidance; customers need to know that you can safeguard their interests and sometimes you will need their help with response implementation.
5. Seek government help when appropriate. Sometimes the magnitude of the problem could be very big and you may lack the resources to effectively deal with it. Sometimes an incident could have wider social implications that are beyond your control. In these types of situations, informing and involving the right government agencies in your jurisdiction is prudent. Many governments encourage voluntary reporting so that industry and society can jointly stay up to date on emerging security threats.

It is conceivable that there will be issues, to build confidence in your products and services, you must be seen as being competent, taking decisive action and communicating effectively. Communication is key because, under threat, it is very hard to deal with uncertainty. Few good practices to remember during such communication are:

1. Inform people (impacted parties and key stakeholders) as quickly as you can. Timeliness is key to any successful communication. This goes a long way in building trust. Bad news early is better than unknown news later.
2. Be transparent and honest in communicating what you know and can share, people, understand you may not have the full picture, especially if you are proactively communicating in a timely manner and that you may have to filter some business sensitive information.
3. Be crisp and clear, ambiguous conjectures reduces confidence.

4. Outline action taken and outstanding plans.
5. Show how you will use this incident for learning and fortifying future defenses. This will shore up confidence in the ability of your business to handle security.

10.11 Summary

Being cyber smart in your products, services and as an IoT solutions company is critical for your success in today's threat-rich environment. In this chapter we have talked about many considerations, tools, methods, and standards that help you deal product and cyber security arising out of multiple threat vectors, data privacy and ownership concerns. Security, privacy, system access and authorization control and compliance are serious topics and will become the cornerstone of your success with IoT. There must be awareness of these issues and engagement for resolution across your entire organizational ecosystem, right from the top. The tone from the top is very important because it sets the priorities and focus. Learning from and partnering with others helps you become better faster; everybody faces similar problems. There are a number of frameworks and services which can help you, but remember your business, your operational context and your threat vectors may be different, so you need to solve these issues uniquely. Engaging customers and other stakeholders in your business are important because they are as exposed as you are. They also may be able to help you get more resources and tools to deal with these exposures. There is a lot of deep thought and hard work required to get this right. Moreover, these issues are as evolutionary as IoT itself. Your policies and practices must be:

1. Reliable in terms of predictable actions and outcomes
2. Recoverable from threats and breaches
3. Adaptable to evolving needs and technology

In the next chapter, we shall discuss other various technology and partner choices to build out your IoT business.

References

1. Dave Shackleford, "A SANS survey", October 2014.
2. Erika McCallister, Tim Grance and Karen Scarfone, National Institute of Standards and Technology (NIST), "SP 800-122 - Guide to Protecting the Confidentiality of Personally Identifiable Information (PII)," December 1990 – present.
3. EC-Council, https://www.eccouncil.org/programs/certified-ethical-hacker-ceh/
4. International Society for Automation, "SA/IEC 62443 - Standards to Secure Your Industrial Control System."
5. ISASecure, IEC 62443 Conformance Certification, http://www.isasecure.org/en-US/, 2017.
6. Mark Clayton, "Stuxnet malware is 'weapon' out to destroy... Iran's Bushehr nuclear plant?," 21 September 2010.

7. Information Security Forum, Standard of Good Practice, "Security management system's usability key to easy adoption," sourcesecurity.com, Retrieved 22 August 2013.
8. Mike Magee, John Lettice and Ross Alderson, "The Register," 1994.
9. National Institute of Standards and Technology (NIST), "Framework for Improving Critical Infrastructure Cybersecurity," https://www.nist.gov/sites/default/files/documents/cyberframework/cybersecurity-framework-021214.pdf, Version 1.0, 12 February 2014.
10. The Stanford Consortium and US National Security Agency, "Research on Information Security and Policy," 1998.
11. UK Government's Department of Trade and Industry, BS 7799, "Information Security Management Systems - Specification with guidance for use," part 2, 1999.
12. The International Organization for Standardization and The International Electrotechnical Commission, "ISO/IEC 27001:2013 - Information technology - Security techniques - Information security management systems – Requirements," September 2013.
13. The International Organization for standardization, "ISO 15408–3.1:2008 - Information technology — Security techniques — Evaluation criteria for IT security - Part 3: Security assurance components," 2008.
14. North American Electrical Reliability Corporation, "CIP 002–009"
15. Craig Gentry, "Fully homomorphic encryption using ideal lattices," 2009.
16. European Telecommunication Standards Institute, "TC CYBER," 2014.
17. National Institute of Standards and Technology (NIST), "Advanced Encryption Standard (AES)," 2001.
18. Ron Rivest, Adi Shamir, and Leonard Adleman, "RSA encryption algorithm," 1977.
19. Partial homomorphic encryption, https://en.wikipedia.org/wiki/Homomorphic_encryption#Partially_homomorphic_cryptosystems.
20. John Leyden, "PC virus celebrates 20th birthday," 19 January 2006. Avoine, Gildas; Pascal Junod; Philippe Oechslin, "Computer system security: basic concepts and solved exercises. EFPL Press. p. 20. ISBN 978-1-4200-4620-5," 2007.
21. Nathaniel Popper, "Decoding the Enigma of Satoshi Nakamoto and the Birth of Bitcoin," 15 May 2015.
22. IBM X-Force Ethical Hacking team, http://www-03.ibm.com/security/xforce/

Chapter 11
Strengthening Your Technology and Partner Ecosystem

11.1 Introduction

IoT is a very rapidly evolving space. This pace and nature of evolution is impacting businesses and companies in very fundamental ways. Large industrial manufacturers are becoming software companies, retail consumer companies are becoming analytics enterprises, cloud software solution companies are developing hardware capabilities that capture more value in IoT ecosystems, start-ups are challenging established players in almost every segment; the transformation across industries is quite unprecedented [1]. Skills and capabilities required to win in this new world order are vastly different to what companies have cultivated over past several decades. Moreover, IoT is impacting transformations much quicker than businesses are normally used to; which necessitates acquiring skills and capabilities swiftly. In such a dynamic environment, it is imperative for companies to partner with others to get faster access to solutions, quicker reach to markets, better scale and more comprehensive capabilities.

From industrial age to information age, we have been debating the merits and demerits of vertical integration of capabilities vs horizontal excellence in them. These approaches defined how we engaged external partners that brought complementary capabilities. The recent IoT age is showing us that success today can no longer be determined by such simplistic models. Companies engaging in IoT today need to operate in a more intricate web of self and partner capabilities that often overlap. So from a 2-axes model of vertical domain specific and horizontal function specific capability model, we have to move to a 3-axes cube model where a partner strategy becomes the third axis. Partners become critical because they may bring speed and scale, two factors we absolutely need to win with IoT. The key issue is deciding where and when followed by how. Because of the evolutionary nature of the IoT domain, we also need the flexibility to move partners in and out with internal or external capabilities. This chapter helps you build the know-how of building and managing your partner ecosystem.

S.R. Sinha, Y. Park, *Building an Effective IoT Ecosystem for Your Business*,
DOI 10.1007/978-3-319-57391-5_11

We shall cover the following topics in this chapter:

1. Defining 3-dimensional capability model
2. Deciding where, when and why partners are required
3. Evaluating technologies
4. Building a partner selection framework
5. Engaging with and onboarding partners
6. Investing in relationships
7. Driving change in your organization to work effectively with partners
8. Keeping your options open—elements of design in technology and services
9. Monitoring partner effectiveness

Technology related and other types of partnerships are not new; collaboration between businesses have existed for centuries. But IoT-driven environments have a few major differences, let us discuss them before we get into the main discussion of this chapter:

1. **Every partner has to bring some technology to the table**—Previously we used to engage partners for their unique knowledge and experiences. Today, no aspect of the business is devoid of technology. So every partner we engage needs to have some unique technology capability that they can bring to our benefit, complemented by shorter response time [2]. For example, you could previously engage a firm which has a specialty in understanding subject areas like user experience. Today if the user experience design firm cannot translate user research and design concepts into a user experience strategy, user interface designs and industrial design prototypes, your objectives will not get addressed. Not only your partner needs to be an expert in design, they need specific capabilities in different types of client devices and display variables.
2. **Technology has to be very contextual**—In the past 50 years, generic technology has been applied quite successfully to solve different business problems. Ability to abstract business layers into basic and broad topics using technology was adequately successful. ERP and other enterprise applications are a good embodiment of this phenomenon. Common processing requirements around standard business process like purchasing, inventory management, production planning were captured in all-purpose databases, workflows and communication models. Even technology embedded in products was not considered key to differentiation; application usage and market channels were used to improve competitiveness. However, miniaturization and deep embedding of technology are changing some of these premises. This requires a very domain-specific contextual understanding of applying technology. For example, the communication needs of a retail consumer device like Fitbit or smart washing machine is very different than those of a home security system which in turn is very different than that of an office building security system. So the electronics and protocols used in these different communication use cases have to be different. Similarly how you identify and personalize an occupant in a home is very different from an office environment; this will essentially define how you do identity management and define user experience in interactive software.

3. **Technology excellence is required from edge to the cloud**—The basic premise of IoT resides in combining massive scale edge computing with cloud technologies in creating transformative experiences and outcomes. Previously companies used to think of applying technology in layers, now more integrated thinking is essential. As we have established earlier in this book, you need to have a comprehensive view of your entire IoT ecosystem, instead of focusing on something which you believe is your narrow core.

Now let us build a model for your technology and partner ecosystem. For explaining the various components of the model, we will use two running examples of IoT-based initiatives throughout the rest of this chapter:

1. Manufacturer of large capital-intensive industrial machinery called Star Trek Enterprises or STE (clearly you can see our inspiration).
2. The maker of a consumer-oriented wearable device called Dangal Enterprises or DE (another movie inspiration).

11.2 Defining 3-Dimensional Capability Model

Once you have defined the model for your IoT-based business, we need to define all the capabilities required for success. The capabilities need to cover the entire spectrum from your product to outcome realization by your customers. You can use the IoT Framework described in Chap. 1 as a reference for listing out the capabilities. Keep iterating the list till you feel it is exhaustive and representative of your actual business environment. While you are building the list, plot all the technical and functional capabilities required on the horizontal x-axis and the more domain specific capabilities on the y-axis. You will find it hard to distinguish between what goes in which axis; this is the reflection of real complexities in IoT-enabled businesses. Often you will have the same thing appear in both the axes, and you will need to differentiate based on keeping the x-axis as generic as possible and making the y-axis as domain intensive as possible.

As an example, for STE, you need capabilities around smart devices, gateways, data transport, data processing, analytical processing, customer engagement applications, validation routines, etc. Similarly for DE, to be successful you need to have a very striking product with contemporary and adaptable design, easy to instantiate and use, inexpensive connectivity, robust aggregation and analytical capabilities run on the scale, very intuitive and appealing user interface applications, etc. While we are talking about two very different domains, you will notice many of the capabilities are similar. In both scenarios, on x-axis you will capture dimensions around industrial design, miniaturization, messaging infrastructure, container design, data storage and retrieval techniques, modeling algorithms, visualization tools, IT application development skills, etc. On the y-axis, you will have elements like environmental factors influencing the design, statutory and certification requirements, data throughput considerations, control algorithms, optimization techniques, value capture and delivery models, etc.

Now you take your partner choices and understand two things—what is their growth strategy and which aspects of the IoT ecosystem are they really good at. For example, Microsoft's growth strategy is based on becoming the best provider of technology processing platforms and tools. Consequently, Microsoft is exceptionally good at their Azure container platform, operating system offerings, analytical and visualization tools, natural language and similar processing engines and programming platforms. Microsoft has many offerings and ways for customers to leverage their technology, so they have also embedded strong cyber security capabilities in their offerings. This is very much in line with how the company has evolved and grown over the past 40 years. Microsoft sometimes creates vertical domain specific modulations of their platforms and tools to make it easier for other companies to adapt them easily. Their strategy is not based on trying to develop competing capabilities and offerings as their customers. The other industry leader in similar space, Amazon, has similar aspirations as Microsoft, but they have the slightly different background and strategy. While Microsoft has historically focused two ends of the usage spectrum—large enterprises and personal computing, Amazon grew rapidly as a very innovative start-up with the strong retailing background. So the analytical capabilities and developer tools of Amazon Web Services (AWS) are exceptionally user-friendly. Their ability to onboard new customers and initiatives is an industry standard. Microsoft and Amazon are very large corporations with expansive capabilities. Even for smaller companies, you can do a similar analysis and understand their strength. But in such cases, it is critical to understand the difference between strong capabilities and strategic growth focus. For example, if a smaller technology provider with exceptional sensing solution could see their primary growth path as an analytical service provider or a sensing technology platform provider, you will have to evaluate carefully on how to engage such a company since their energies and investments will be in different directions.

Once you have a listing of these various capabilities—internal and partner sourced, you may want to plot them on a 3-dimension model for easy reference, usage, and updates. It will also be helpful to identify the capabilities you have or need, which are or have to be industry leading and which ones will bring maximum differentiation. Figure 11.1 shows a representation of how to think about this 30-dimensional model.

Remember a few things about building and more importantly maintaining such a model:

1. This is not a static model and will evolve with time, technology, business models and partner choices; so frequently (at least every 3 or 6 months) update the model
2. Greater granularity makes it more meaningful, but you may choose to use different abstractions for different audiences
3. Determine internal capabilities through a rigorous and objective process
4. Ascertain partner capabilities through a transparent dialogue process
5. Evaluate your competitors' actions and use the insights to improve your model

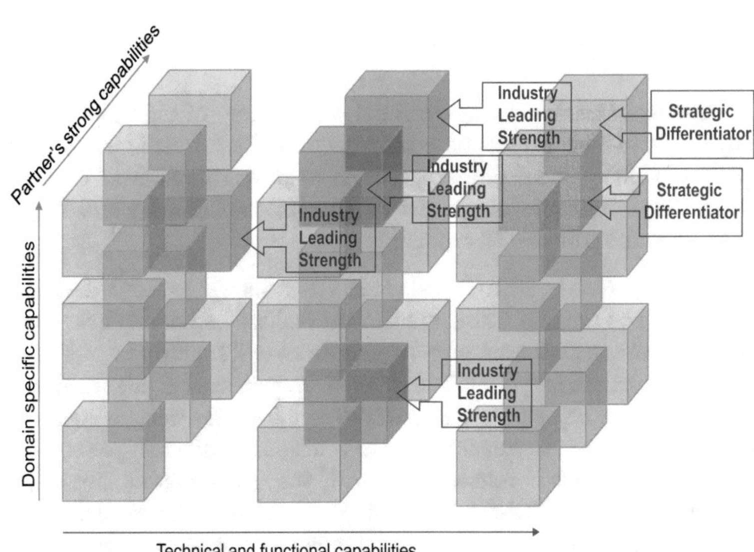

Fig. 11.1 Three-dimensional capability model

The last bit around competitors' action is crucial because you will have to constantly monitor the threats to your business, understand the changing landscape and develop appropriate responses which need to be rooted in your capabilities. You need a high degree of paranoia and wide scope view to keep winning in IoT dominated businesses since the entry barriers are somewhat low.

11.3 Deciding Where, When and Why Partners Are Required

In the traditional technology space, you can reach out to one or two large companies to solve all or most of your problems. If you are trying to redefine customer management process, you can either enlist a reputed business/IT transformation consultant or more recently reach out to somebody like Salesforce.com; for a financial transformation project, you can talk to the likes of Oracle, SAP, IBM, Accenture, etc. Unlike other technology domains, in IoT, no single player has emerged with a comprehensive solution set—none whom you can go to with an idea as a one-stop shop to build your business. You will have to either build nearly entirely on your own or carefully string together a solution ecosystem with technology components from many. If you are trying to build IoT businesses quickly and effectively, partnerships are a viable and sometimes necessary path. Moreover, customers today care less about what you do yourself, they care more about how you solve their problems.

11.3.1 Partner Classification

It is challenging to neatly classify potential partners into specific categories as we normally would in other aspects of technology or business functions. This is due to the rapid evolutionary nature of this space and adaptive aspirations of these companies. But to help us navigate this exercise, we can use the following broad categories to classify potential partners for evaluation:

1. **Edge device**—Nearly all IoT businesses have 'things' or edge devices. Edge devices could be smart or made smart. We have discussed smart devices at length in an earlier chapter. This category of companies will supply you with devices or technology required for your IoT business or help make your edge devices smart. This spectrum of companies will cover everything from design, identify material, source and assemble/manufacture to your specifications. If you are an established business which is venturing into IoT, you may prefer to have an original device manufacturer (ODM) relationship with a partner. For smaller companies, white-labelling somebody else's product might be a better strategy.
2. **Gateway**—Companies which help your edge devices connect to the internet or cloud by providing gateway solutions will fall into this category. Gateways could be independent boxes or small processing units that get embedded into existing edge devices. A large number of chipset manufacturers are expanding their portfolio into gateway devices. While it is perfectly okay to choose an existing gateway solution, evaluate the cost and functionality fitment for your business and use-cases.
3. **Communication**—Companies which help with data transport will fall under this category. Cellular or wireless communication providers like Verizon, AT&T, Vodafone, Orange, Comcast, etc., all will fall in this group. You will also encounter companies who can help you design transport solutions using existing networks—Wi-Fi or LAN or otherwise; such companies will also be classified as communications companies.
4. **Data platform**—This is a very broad category of companies who help you with storage and processing of data. In our IoT Ecosystem Framework, all activities under Device Management, Data Routing and Real-time Analysis, Data Service Platform, and Analytics will fall in this group. You will encounter cloud container solutions (and often more) providers like Microsoft and Amazon, large analytics platform providers like GE Digital and IBM Watson, a whole range of smaller start-ups like Predii, hybrid companies like Thingworx [3], and many other variations. This is a very crowded space and rapidly growing every day. Very rarely companies in this category will help you with the previously described upstream steps, but they will often try to engage with you on the upcoming downstream stages. Almost all these companies will promise a virtual plug and play capability with access to their vast analytical and learning capabilities in the platform. The promise and pitch are usually very attractive, but you have to evaluate several things when engaging with such companies—your long-term lock-in, cost structure, efforts required to adapt their platform to your business needs.

In our experience, depending on your team's maturity and capability, you may find it quicker-easier-cheaper to build the data platform. But our recommendation is also not to attempt to build your own cloud container solution because it can be cost prohibitive and capability limited compared to the offerings of giants like Microsoft and Amazon.

5. **Delivery**—This category of companies will also fall into a broad range of capabilities ranging from ones offering professional services to build your delivery applications, providers of visualization solutions, or boutique firms with specific capabilities in the business delivery activities.
6. **Business support**—Many companies which evolved from traditional enterprise solutions base and now are expanding their IoT aspirations will fall in this space. You will also find some boutique platform solutions providers, mostly start-ups that operate in this space helping you with billing, CRM, partner management, etc.
7. **Security**—There are some specialized companies working with security aspects of IoT businesses which will fall in this bucket. Such companies can be your consultants, security developers, intrusion analysts, validation providers or ethical hackers. We are yet to see large comprehensive security solution providers with a niche focus on IoT. Depending on the maturity of your internal cybersecurity practices and capability of your in-house cyber security team, it is always good to evaluate engaging with an external security partner.
8. **Specialty**—There are a number of unique services that do not fall into any of the specific categories from above; all such companies will become part of this category. Specialized product or user-experience design firms, research, and marketing firms, advisory firms specializing in the financing of IoT businesses, etc. will fall in this category.

11.3.2 Engagement Drivers

Based on your internal capabilities and drivers like time-to-market needs, logistics, interrelationships between selected partners, physical co-location, etc. you will choose to engage with external partners. Here are some guidelines to help you think through when to engage with such companies:

1. You are just starting and have limited capabilities in a specific part(s) of the ecosystem. For example, your manufacturing capabilities are very good and can be easily extended to make smart devices, but your experience with data analytics is very limited; in such case, you may want to engage a data platform partner. In another example, you may have good capabilities in edge devices and data platform, enough to get your business started and scaled to a point, but have limited capabilities in device management; in such case, you may want to solicit a partner with device management solutions.
2. The scale is critical and you cannot reach scale in a reasonable amount of time. This is the same argument often used to making acquisitions. Building scale is often a physics problem—function of mass and momentum. Irrespective of your

willingness to invest, there is a time dimension to building scale, in such scenarios, partnering is a better option.

3. You decide that part of the ecosystem is not cored to your business. While you may have capabilities in every dimension of the IoT ecosystem, you can strategically make choices to focus on what creates core value for your business and customers. For example, as a large industrial IoT company, you may choose to excel in devices and data platform, but leverage partner capabilities for business delivery and support. Similarly, as a consumer-oriented IoT company, you may choose to win with your data analytics and business delivery capabilities, and decide to partner for edge devices and device management. Remember, irrespective of how big and resourceful you are, there is only a limited number of things you can get very good at in a reasonable amount of time, so choose wisely.
4. It is cheaper to use partner's capabilities/platform than your internal efforts. As mentioned earlier, we have seen that using somebody else's hosting infrastructure is usually a better economic decision. Similarly, you can extend this logic to almost anything which is not absolutely core to your success that you must own as technology IP and capability.
5. Technology is changing too fast and you do not want to get locked down into any specific option. This is very true for gateway and communication dimensions of your IoT ecosystem.
6. You get access to a bigger ecosystem by partnering. Sometimes there are opportunities to learn from other friendly or non-related industries. This can be true for almost every dimension of your IoT ecosystem. This is especially true if your maturity in the IoT space is lower than aspired.
7. Your partner helps you get access to new customers and install base. Usually, in such events, you will have joint go-to-market opportunities or receive marketing support for your offerings. This phenomenon is most observed when platform companies with large investment bandwidth are trying to leverage partner base to promote their technology.
8. Your budgets are limited but the partner is willing to invest. A number of start-up companies benefit from this trend. If you are just getting started and lack resources, this method is an obvious choice. Even for larger companies with limited resources, they can explore a three-way partnership with a start-up and a venture capital investor to build out an IoT business. We see this trend picking up in some markets.

11.3.3 Engagement Models

There are a number of different models in which you can choose to engage with your partners. All of them are established business transaction models, but you may want to incorporate specific flavors related to your IoT business into the engagement.

1. **Supplier model**—In this type of relationship you are buying goods and your partners' commitment is limited to quality, schedule, cost and such parameters. You will typically engage in this model with your hardware and electronic component providers. You may also buy off-the-shelf software in this method.
2. **Services model**—When buying services you are augmenting your capabilities through external resources. Since IoT is still maturing, very few services companies have broad capabilities; so you may want to vet out the individuals for their expertise and experience.
3. **Consumption model**—This is similar to utility based pricing where you pay as you go for the amount of technology you are consuming. Typical arrangements in this model are prevalent with cloud platform providers like Microsoft and Amazon; they started with just hosting services being offered through consumption-based pricing, and now are extending to more and more software offerings. This is a good model to reduce risk, manage contractually, modulate growth, achieve scale, and have limited liabilities. On the other hand, you will have to keep yourself up to date on all the changes your partner is making to the platform to ensure your business is still compatible.
4. **Sharing model**—This can be very attractive for the engaging parties as there is an element of risk-reward-investment built into the design. However, this can also be challenging to construct due to the huge number of unknowns. Sharing models are typically much longer term than others because parties making the investment will plan on recovering the returns. Typically technology providers with deep pockets and huge growth ambitions will want to engage with you in this model. When designing such a model, be careful not to compromise on your long-term prospects, and most importantly your access/ownership for customer relationships and data. You will also need to carefully think through all the IP implications and ownerships. We recommend entering into a sharing model only once you have reached a certain level of maturity in your IoT business.

There is no one right model and you may have multiple different ones with your partners. Like any other contract, it is good to clarify the drivers, objectives, and outcomes beforehand along with set governance rules.

In any event, while engaging in partnerships in the IoT space, please be careful of the following:

1. If you are working on cutting edge technologies, chosen partner must have proven track record of working with new technologies successfully in the past. There are a number of capabilities around innovation, rapid prototyping, agile disciplines, etc. which must be matured in the partner to help you.
2. Never get too dependent or fully locked into a specific technology or partner space, IoT will continue to evolve for the next several years before clear winners emerge. Your ecosystem architecture, solution design, and business model need to account for this flexibility, more on that later in the chapter.
3. Constantly evaluate the economics of your offering and service delivery to remain competitive. Prices for electronic components and hosting are coming

down fast. Pricing models for various types of technology and services are also changing. So keep looking out to see the market evolve.

4. Your partner should not become a potential competitor or player in what is core to your business. IoT is blurring the lines between different industry segments and companies' offerings. You should retain IP and platform ownership of what is core to your long-term growth strategy. Often companies make the mistake of depending nearly exclusively on partners for analytics and data platform; we believe this is detrimental in most cases.
5. Limit the number of partners you engage with. There is not enough time to develop the platform and the business on it, the ecosystem also becomes very complicated with too many partners, and eventually economics of the business will get impacted adversely. So invest your time and money with few, but go deep to get best returns.

11.4 Evaluating Technologies

Evaluating enabling technologies is one of the most critical steps in building your IoT business. It is a very involved and often complex activity—there are too many choices, space is evolving too fast, and every solution provider has a very different perspective on how they are the best and will change the world. Often you will feel limited by your internal capabilities to make appropriate choices. There are not too many good advisory firms specializing in IoT who can help you either. So you will have to end up first building a small team that can lead this exercise. Your next step is to definitely create an outline of your IoT ecosystem. We have provided a reference framework in this book, use it to build/adapt your own. You may want to engage some good academics (institutions or individuals) to augment your internal capabilities; academics bring a broad and different perspective, they have a deep appreciation of the underlying technologies, and are usually connected into the mainstream networks of what is going on. A good way to learn is also from your potential vendors. In order to build a relationship and secure your business, they will share a lot with you. Once you talk to several similar providers, you will start to identify patterns around possible usage.

When exploring technologies, there are a few standard questions you must evaluate them on:

1. **Does this fill a critical gap in your IoT ecosystem?** When asking this question, you must make a build vs buy decision. If you choose to source the technology, try to get a bigger functional scope than a very narrow niche. For example, if you decide to use somebody else's data service platform, get one that can process structured, unstructured and streaming data, just not one which has a simplistic Hadoop-based solution to deal with unstructured data. At this stage, you need to extrapolate the various complexities your IoT business will encounter and how technology can be applied to solve them.

2. **Does this interface easily with the rest of your IoT ecosystem?** You cannot afford to have your IoT ecosystem look like a technology museum that requires an army to maintain. Easy interfacing with other technologies in your IoT ecosystem is a key criterion for selection. For example, you choose to leverage somebody else's device management solution in your platform. However, if this requires you to have a complicated and unnatural way to recognize customers, assets, and transactions, this choice will create huge overheads for processing in every other part of the ecosystem, rendering this choice a less desirable one.
3. **Do updates to this technology seamlessly integrate with your IoT ecosystem?** This consideration is an extension of the previous one. Sometimes technology might be easier to implement and initially integrate but might have an upgrade path which is quite challenging. This could happen due to technology or component obsolescence or simply a very different direction taken by the provider. While one can expect IoT technologies to rapidly change, but it that cannot unsettle your complex IoT ecosystem. For example, you can choose a security implementation from a partner that uses very complex software based encryption and authentication methods. If their roadmap is to embed this capability in a proprietary hardware solution, that might cause major issues for your implementation. To avoid such issues, you need to study your partner's roadmaps very carefully and participate in deep conversations around technology evolution.
4. **Is it relatively easy to replace this technology with another alternative at a later date?** This is a tough consideration, we will admit that. While you want to retain as much flexibility as you can, you do not want to use too much of consumer grade technology to build some differentiation and competitive advantage. But try and avoid any technology which can diminish your customer and channel access in future. You have to be very careful on these lines when making choices around electronic components and communication techniques, it is very hard to change those choices later because they become part of the core infrastructure of your offering.
5. **Does this technology create any major disruption or disadvantage for your customers?** Sometimes your choices will cause inconveniences for your customers limiting your offerings' adoption. Imagine if a Fitbit could not handle a little bit of moisture and heat, consumers could not have used it during exercising which is one of the key usage times for such devices. Imagine an industrial equipment manufacturer has created a new path breaking predictive diagnostic capability using vibration sensing but requires zero acoustic disturbance for such sensing to work. To make this work, the customer will have to make significant investments in structural adjustments to the housing of the said industrial equipment, which might be too complicated and costly. In another example, a particular solution might require all kinds of personal data to be stored and accessed easily on the cloud which might be beyond the comfort levels of the users. You need to understand all such implications before making your choice of technology.
6. **Does it provide you with a distinct competitive advantage?** You want technology choices that can help create better customer stickiness. You can also explore denying your competitors access to your chosen technology, at least for a period

of time. For example one IoT business has chosen to use Windows operating system for the device, Cortana as a user experience engine and Azure as the data platform. This helped the company develop its solution with comparatively fewer efforts and quickly as Microsoft has worked on interoperability across its platforms and built-in security across the processing chain. For customers who are naturally Microsoft users, this solution will be appealing and easier to adopt. In another example, Nike has embedded sensors in its shoes to track activity and then integrated it with a smart sports watch it makes. This creates a seamless experience for athletic or activity oriented customers, enables unique appeal for both the Nike products and creates barriers for competitors in both categories (shoes and smart watches).

7. **What are the next best alternatives?** It is important to assess the next best alternatives while evaluating technology choices because it tells you a lot about category cost structures, potential competitive differentiation, ease of implementation, current competitive usage landscape, and gives you an idea about probable evolution paths.

Convert these questions in a logical flowchart that is most appropriate for your context to complete this evaluation.

11.5 Building a Partner Selection Framework

So far in this chapter, you have been exposed to some tools for identifying your capability needs, understanding partners and exploring technologies. Now you need to work on ascertaining the partners who will help you build your IoT business. Following are some key criteria to evaluate potential partners:

1. **DNA**—Every company and every solution has a core, something which they are exceptionally good at, something around which their entire existence has been designed, something that uniquely defines and differentiates them. While everyone tries to expand their capability and offering portfolio, they will always be best at their core. For example, Salesforce today has a whole range of enterprise offerings through its partner network, but it is still one of the best choices for sales automation and analytics; Apple will always be one of the best user experience company; Microsoft will always be one of the best productivity solutions company; GE will always be one of the best industrial solutions company; Johnson Controls will always be one of the best building controls and HVAC companies; 3M will always be one of the most innovative consumer solutions company; and so on. To understand the DNA or core of a company you need to understand what and how they evolved. Then you need to understand other aspects like is their business model based on scale (mass market) or uniqueness (boutique), this will indicate how you should leverage your partner solutions. Finally, you need to gauge your partner's focus on continuous improvement and cost reduction to judge how your solution will evolve in this partnership.

2. **Technology leadership**—It will be very challenging for any company in the IoT space to be successful without technology leadership. You need partners who are on a similar trajectory. You need to find out if your partner has the right focus, desire and capability to be on the leading edge of technology. This is demonstrated through a number of things—deep understanding of the core technology, appropriate R&D investment levels, quality of technical people, IP, peer level and diverse industry acknowledgment, academic associations, solution depth and the overall tone of the company. Awards and high placements in things like magic quadrants are good indicators but cannot substitute the other factors mentioned.
3. **Competitive threat**—There are many aspects to competitive threats—the industry you operate in is moving towards a common technology footprint that could isolate you, your competitors are making some technology choices that will create a distinct advantage for them, your competitors are making the same choices as you but are investing more and executing faster, your product/offering category is getting obliterated by a different substitute. Each of these presents a different problem domain and require a distinctive response. You will need to arm yourself with technology and partners that protect your product and your position. Taking a leadership position with technology and aggressively promoting such choices in your and adjacent industries is often a good ploy, but you will need the patience and investment for that. Alternately you can also ride on the coat tails of giant technology suppliers with a lot of resources to achieve similar goals. By deeply integrating your domain expertise to technology, you can ward off category disruption threats. Sometimes you will want your partners to have a deeper relationship with you than your competitors. It is unlikely that you will enter into exclusive relationships, but you can surely get greater mindshare.
4. **Capability**—You need to consider your potential partners' capabilities (people, tools, infrastructure) in terms of how it complements yours and is comprehensive enough for your needs. Sometimes people find these two factors are contradictory, but they are not; you want your partner to step in nearly every activity area should the need arise. A good way to review partner capabilities is using the same capability matrix we talked about earlier in the chapter.
5. **Consultative approach**—Most often you want a partner who can collaborate with you instead of just reacting to your instructions. Given the evolutionary nature of IoT and endless possibilities, you want a joint discovery journey with your partner. However, to make it more structured and shorter, it will be better for your partner to bring some frameworks and past experiences, so look out for those. You definitely do not want generic management consultants or technology services providers who have neither context of your industry domain nor specific knowledge of your technology ecosystem, you want people who can really take you forward with empathy.
6. **Economics**—This is a very important consideration because in spite of all novelty factors IoT is a very competitive space. So you need to keep your technology cost and development cost as low as possible. Also, it is critical not to have long-term lock-in cost, you want maximum possible flexibility in cost structure. Since

this is not a matured space, look for the total cost instead of unit or transaction costs, sometimes seemingly cheap becomes more expensive.

7. **Long-term viability**—IoT is flooded with start-ups and companies that are tipping their toes to ride the hype-cycle. For many such companies, their goal is to create some quick wave and attractive valuation by capturing some install base of their technology and services. You are in for the long haul and you want partners who can work with you for a long time. So you need to evaluate the long-term viability and interest of your partners. There is no easy and obvious way to make this judgment, you will have to engage in deep conversations to understand your potential partner's long-term goals, business model ideas, cash flow and reserves, current and future expected valuations, exit plans, and investment profile.
8. **Independence/Reliance on other partners**—Often you will see companies building out capabilities and offerings entirely on their own or leveraging a larger network of other partners. This is very common in any emerging technology space. Depending what you are trying to achieve, both options can be good. You have to be particularly careful when evaluating electronic components and communication partners because they usually have a dependence on other suppliers or broader adoption networks. For example, if you have an edge device or gateway partner who is not associated with companies that have very large applications of chipsets and communication modules, you will be cost and technology disadvantaged in the long run.
9. **Compliance**—There are a number of different agency and compliance certifications required for IoT products. It is critical for your partner to have exposure to the ones that are relevant to you in order to shrink cycle time for development and launch.

You need to create a formal evaluation matrix with weightages for each of the above factors tuned to your business context. Involve internal people with diverse background and perspectives to do the evaluation in a collective effort.

Once you complete the evaluation process, explain to your partner the rationale behind their selection along with your expectations. Transparency and alignment on objectives is key to the success of any future potential partnership.

11.6 Engaging with and Onboarding Partners

Now that you have selected the partners who will join in building out your IoT business, you need to thoughtfully get them started. You cannot expect to throw a statement of work across the wall and expect phenomenal results in return that help you win. Building out your platform and business has to be a cooperative process. There are a number of steps you should take to make this a meaningful partnership. This is an iterative exercise. Your partners may not have enough background in your business; it will take them a while and several rounds of education to understand the intricacies of your environment. You need to be open, share as much as you can without compromising your competitive edge. Remember once a trust-based relationship is established, you get back as much as you put in. Given the time sensitivity to any IoT initiative, effective partner onboarding is key to shortening cycle time; in our reviews of various IoT businesses, we have been able to trace back a number of downstream issues to lack of understanding upfront.

1. Give your partner an overview of the following in the context of your business:

   - **Economic value chain**—Explain to your partner how money moves in your industry or product category and how value gets created at each step. Porter's Value Chain Analysis model from his book *Competitive Advantage* is still a good tool to start with. The economic value chain analysis is useful to understand what customers' value and how much they value it for, and the cost build-up for creating the perceived value. In the case of regular good and services, the value chain usually has a one pass flow; in the case of IoT businesses, there is iterative value creation over time and usage because data over time creates new forms of value. For example you buy a car and are willing to pay a price for it one-time which eventually diminishes over time; however you buy a predictive maintenance or instrumented insurance service, you will continue to pay for it over time as it accrues more value for you. If you understand the economic value chain well, you may choose to price the predictive maintenance or instrumented insurance services to increase in a progressive manner as your need for them rises with time. This allows you to create newer value streams over time.
   - **Information value chain**—Clarify to your partner what and how information gets created in different stages of the processing chain, their implications, how to interpret them and what value it can create for rest of the flow. This is similar to the economic value chain but deals with data and insights. This exercise should be done iteratively with increasing depth. This goes on to explain the significance of data to each aspect of your business.

2. Conduct detailed orientation of your IoT business:

   - **Ecosystem framework**—We believe that each IoT business will have a slightly different ecosystem framework to capture the essence and differentiation. Your partner needs to understand the basic infrastructure design of your platform enabling the business. Explain the flow of data, insights, and value in the context of the ecosystem framework. This will help you and you partners to have a common reference point and language for future activities. Keep in mind that your ecosystem and platform has elements of competitive advantage that you want to protect, so share accordingly. For example, you may have a unique way of processing time-series data or managing security that puts you above the rest, do not get into the details of those; just limit the orientation to overview.
   - **Offering framework**—Offering framework is how you choose to deliver your business to your customers and capture value. This will look very differ-

ent from your ecosystem framework and reflects your monetization strategies. Your partners needs to understand the offering framework because it allows them to put in context the economic value chain analysis and what outcome the IoT technologies must instantiate. The offering framework like the ecosystem framework and underlying technology platform will evolve over time; an initial orientation creates a perspective for future changes.

3. **Partnership framework**—It is likely that you have multiple partners. It will be useful for each of them to get acquainted with others, understand mutual roles and responsibilities, and have some common rules of the game to play to win. Sometimes the boundaries between various partners and technologies will not be clear. Moreover, every player has aspirations for portfolio expansion and volume growth. For example, your data platform partner will want to have a slice of the delivery and business support activities; device management and data platform management are converging; and so on. In order to get more mindshare, techshare, and business, your partners may get misaligned with your end objectives. Establishing the rules of competition early on and keeping them transparent is crucial for your success.
4. **Do a pilot project**—Nothing furthers the induction process as doing a hands-on project by the chosen partners. This allows you and your partners to understand each other's working rhythm, operational processes, execution language (like a development process), and build relationship amongst teams, perspective about quality and cultural facets. A shared success or shared learning will build a strong future foundation. The pilot project should not be a throwaway; it should be a problem you are trying solve or a component you are trying to build. It should be sizeable enough to give everybody exposure, but not too big to jeopardize the bigger effort. For example, with an edge device partner, you could just develop one smart edge device instead of an entire line card; with a data platform partner you can work on ingesting data from one source, doing some device management, running a couple of algorithms, identifying some insights and tying it to some minimal customer value.
5. **Identify shared success metrics**—You and your partners may have different motivations and reasons to enter into this endeavor. Be clear upfront about your drivers, common meeting points and how will you ascertain success. This has to be done on a longer time horizon with intermediate definitions of success. For example, with a data platform service provider, your common goals might be platform usage metrics because it drives consumption for the provider and potentially bigger landscape of opportunities for insights for you. Now increasing consumption does not necessarily drive greater revenues for your partner because the cost of storage and processing is coming down and you may be continuously optimizing data management. If key success criteria, in this case, are not clear upfront—consumption or revenue or both, both of you may pursue different paths. Similarly, your partner needs to recognize that they can keep bringing newer capabilities in a consumption model to increase their revenue opportunities, but unless they create directly attachable monetization opportunities for

your business, your proclivity to consume these new capabilities will be very less. So you need to have shared an understanding of innovation outcome as well.

6. **Establish IP generation and ownership framework**—In the IoT space, intellectual property (IP) gets generated in many different ways and stages of the business, definitely at a higher pace than other businesses. You will create IP in distinctive usage of preexisting algorithms, how you create insights in a specific domain, how you manage devices and communication in a specific industry or device class, how you manage data and your platform, how you implement security, and the list goes on. You can literally create dozens if not hundreds of IP for each dimension of your IoT ecosystem framework, we have personally seen that happen in our work-life. The dynamism and optimism of IoT fuels this culture of innovation deeply. You will always want to keep all IP with your organization. But make sure you create enough incentives for your partners to contribute to the process of innovation. Also, have mechanisms in place to recognize and reward any outside innovation your partners might bring to help your business.
7. **Determine data ownership and rights to use**—Data often becomes a sticky point in most IoT businesses. Customers will assert their right because of ownership of the final product/service; you will try to assert your rights as the one who is enabling it; your partners will try to claim a piece of it as they might have helped you access or use some of the data. Everybody believes that ownership of data will define your economic benefits from derived insights in future. We have seen companies engage in otherwise financially non-viable transactions just to get access to more and diverse data (which is not a good practice for IoT or otherwise as it goes against business fundamentals). Be very clear about the relationships around data ownership and more importantly usage with all your partners and stakeholders. You want to avoid two things—being denied access to the data in future, and enabling another party to offer in future the same services/product/capability that you are bringing to your customers today.
8. **Resolve exit criteria and process**—Ending partnerships is a normal business process. In the context of IoT, it acquires some different connotations. You may be sourcing technology or components from your partner that you will need either access to or be supported for a longer period of time than the length of the relationship; this is very applicable in most industrial IoT cases. It is also possible that both you and your partners have access to data from your customers which you can apply in different ways in future to disadvantage each other. There are always tricky liability issues around IoT because of the number of transformations that data, insight, and actions undergo throughout the processing chain. In all cases, you must have clearly identified exit criteria and post-exit roles and responsibilities between different parties.
9. **Define partnership governance process**—Governance is an indispensable element of any successful relationship, no different in IoT. The governance process should outline mutual expectations, success metrics, celebration models, review process, escalation process, key personnel and cadence for various governance activities. It is important to identify the celebration models and ideas because

you are both venturing into something new and exciting, you want to tell the whole world about your success and create a self-reinforcing cycle.

11.7 Investing in Relationships

No successful relationship is a one-way street; this has been known and practiced for a very long time. In building IoT businesses, we have less time and increased reliance on our partner ecosystem. So our necessity to invest is more severe. Here relationships have to be based not only on trust and mutual respect, you also need a much deeper insight into each other's organization and capabilities; this will allow you to draw upon each other's expertise quickly.

In our view, there are three key pillars to any such successful relationship:

1. **Relationship manager**—Two organizations come together because they see the value, but the relationship managers realize that value through relentless passion, commitment, and advocacy. Through your other business dealings, you have come to expect to deal with account managers who are primarily motivated by sales and customer satisfaction; to reciprocate you will also have a designated vendor management organization or person. Relationship managers act beyond that, they are personally invested in shared success. They must also have a deep empathetic understanding of the other organization including structures, policies, politics, and possibilities. A good relationship manager will be skilled in navigating both internally as well as with the partner/customer. As the business owner, you must reciprocate partner efforts with your own relationship manager. Depending on the depth and critical significance of the partnership, you may need senior level people to champion the relationship.
2. **Reviews**—Meaningful and periodic reviews are an integral part of any successful relationship. Your IoT partner reviews need to be a bit different than your normal partner reviews because space is rapidly evolving (hopefully your business is too), both of your survival depends on shared success in this new space, and technology is more relevant here. There are many dimensions which you should review together with your partner:

   - **Landscape review**—This includes talking about how the IoT space in general is emerging and evolving, what are the new trends, who are the new entrants, who is exiting and why (tells you about possible failure or exit scenarios), what kind of new technology is becoming popular, what type of new business models are appearing, new standards and practices, regulatory implications, and analyst predictions. Your partners may have wider access to all of these happenings in the industry which can help you learn and design strategies for your success.
   - **Business review**—This part should focus on how each of you are making progress against your plans and aspirations. Talk about revenue and customer growth, new offerings and capabilities, challenges and failures experienced

with lessons learnt, continuous improvement efforts and results, new initiatives and dropped initiatives, marketing efforts, new IP, and other topics which might be useful for both.

- **Program review**—This is similar to a regular program review with deep focus on technology and execution. You can expect many variations in implementation of technology and how you expected to run your initiatives; discuss the observations and lessons learnt with your partner to influence your future course of action.
- **Personnel review**—Success in IoT has a huge dependence on working with the right set of capable individuals, more than other fields, because expertise is novel and limited. So you should talk about the rising stars in both the organizations to explore better ways to recognize and leverage them. You should also talk about rising stars outside and if there are ways to tap into their expertise.

3. **Executive connects**—IoT presents many transformative opportunities. Often it requires a much higher level view and abstraction to comprehend the possibilities. This will allow correct resources to be deployed by you and your partner companies to capitalize on the possibilities. This is best achieved through executive connects because it is easier for such a forum to make those connections. If the future of your business is dependent on the success of your IoT initiatives, making executive time available for such reviews is important. Instead of trying to present only high level summarized updates, engage in a few in-depth conversations on the lines of topics outlined above; executives will appreciate the extra education and get more involved purposefully.

11.8 Driving Change in Your Organization to Work Effectively with Partners

By now you have been exposed to how IoT brings differences to your normal business practices and even how you manage your partners. In normal situations, we talk partners but think vendors; in the case of IoT, we have to think and act partners for speedy success. So you may need to change some customs and behaviors to work effectively with partners. Driving any such change has to be thoughtful and deliberate, integrated into the operating rhythm of your company. We have discussed many changes throughout the chapter, here are is a summarized list for you to keep in mind:

1. Share everything (without compromising any confidential or highly strategic competitive information) about your markets, business, plans, products, customers and value chain that is necessary for success. Withholding information here is not a source of power but a cause for ambiguity.

2. Focus on your partner's success and not only yours'. You need them as much as much as they need you, so let them win too. We have seen too many procurement organizations trying to squeeze the partners such that their interest diminishes dramatically.
3. Explain to your staff the partner evaluation and selection details, it will help them understand better why they need to pay attention to the partner.
4. Encourage your internal staff to learn from your partners and just now leverage their services or solutions. This will help with your self-reliance, make your employees more valuable, and develop a sense of security. This will also enable you to leverage your partner capabilities better and have a healthier mutual respect.
5. Create multiple joint work-a-thons, folks from your team and your partner's team working together in a somewhat unstructured manner. We get very focused on project tasks and plans under pressure to perform; without such forums, it is difficult to build a good joint working rhythm. There is a lot unknown still and together we can explore more.
6. Collect feedback constantly how things are moving forward so that you can address any issues timely, you are in a race against time and do not want to lose any time in friction.
7. Support your partner in their promotional and development activities. Remember, they are also trying to build a business like you and they will likely remember their friends who helped them get there.
8. Use any partner forums and trade shows to socialize and learn from other companies like yours'.
9. Do not discourage your partners from working with your competitors, after all, there might be something for you to learn too.

11.9 Keeping Your Options Open: Elements of Design in Technology and Services

While partnerships are critical to your success, you also need to be self-reliant. Earlier in the chapter, we have briefly touched upon why you do not want to get locked down with few choices. The issue here is how; in order to be quick, you want to reuse as much as your partner has to offer, but that reduces self-reliance. Because there are no monolithic technology solutions (at least yet) in IoT, you can use that as an advantage in your architecture and design to reduce reliance on partners. Following steps help you accomplish this:

1. Layer technology as much as possible. The primary objective of developing the ecosystem framework was to identify the multitude of layers that exist in any IoT environment. If you separate the design and implementation of these layers to as

much granular level as practical, you can make different choices at each layer. For example, keeping your messaging and transport layers for data allows you to engage two separate partners and swap them in-out as necessary. Similarly separating hosting containers from analytics will allow different solutions to be leveraged for both. Keep storage, processing, analytics, messaging, device and platform management, delivery and business support layers as separate as possible.
2. The interaction between layers of your ecosystem should be through public services instead of being hardwired so that when you make different choices later, you have only limited places to touch and change.
3. Abstract business logic in what is your own. Embedding your business deep into your partner's technology increases reliance.
4. Use agent-based models in design. Agents allow for the most miniaturized microservices in a complex computational environment. This approach helps you design in most optimal modular and scalable manner.
5. Leverage open source and generally available technology as much as possible. It reduces reliance and cost. This allows you to tap into bigger resource pools for development and support. It also helps improve security and stability of chosen technologies as more people contribute to the hardening efforts.
6. Simulate chaos monkey exercises. These are resilience exercises to evaluate how badly you will be impacted if random events occur. This will help you design recovery and self-reliance programs.
7. Have your own team capable of doing almost everything. Even if your internal team does not have the full scale and capability, you should have a doomsday survival scenario played out.
8. Periodically evaluate alternate technologies and suppliers. Keep current with what is going on, it will provide you with direction to make different choices.

11.10 Monitoring Partner Effectiveness

When you lay out your IoT strategy, there are a number of objectives you arrive at for the business, some of the common ones are—revenue and profits from new connected offerings, market share growth, access to new market opportunities, penetration into existing (if applicable) install base of assets and customers with connected offerings, and new value streams captured. Your partner ecosystem in which you are investing so much is expected to help you achieve these goals quicker and better. So you need to build a framework and process for evaluating partners that are simple yet effective. Below is an example of such a framework. These are some suggested categories and measures to get you started, you should design the framework that best suits your business.

Category	Measure	Weightage	Score
Financial impact to your business	Revenue from offerings that your partner had a significant role in building/enabling		
	Revenue from new innovations identified or developed by your partners		
	Cost reduction or margin improvement in your offerings and operations through partner efforts		
Speed to market	Amount of time saved by partner technology/offerings		
	Number of new market channels your partner has enabled access to		
	Number of partner solutions/offerings you have been able to largely repurpose to grow your business		
Competitive impact	Number of competitive barriers enabled by your partners		
Technology leadership	Number and impact of new technology innovations fostered by your partner		
	Number of new technical capabilities introduced by your partner		
Mutual reliance	Your partner's share in your IoT ecosystem		
	Your share in your partner's business portfolio		
	Number of your offerings where you are completely dependent on partner capabilities		
Promotional impact	Number PR instances and platforms enabled by your partner		

In doing this evaluation, please note a few factors:

1. A high score does not always indicate better. For example, if your offerings are too dependent on your partner capabilities, you have a higher risk.
2. Weightages can change over time as the context of your business and objectives from your IoT efforts evolve.

It is always recommended to share these findings with your partner in a respectable and transparent manner so that both of you can take steps towards improvement.

11.11 Summary

The journey to building an impactful IoT business is challenging. You need partners to help you in different aspects of building out your ecosystem. If you are a large corporation, you need nimble players to help you capture opportunities; if you are smaller and have limited resources, you need others to help you compensate; if you are a start-up riding the IoT wave, you need customers and applications to see your solutions actualize; in any scenario, you need partners. Businesses have been historically built through collaboration between multiple entities; in the space of IoT partnership acquire a new meaning because of mutual reliance for success.

In this chapter, we have exposed you to why you need partners, how to find them, how to leverage them and finally how to keep your options open should something change. Building effective partnerships is not a trivial task; it requires deep thought and hard work from your end.

In conclusion, please remember a few closing thoughts:

1. Working with partners gives you easy access to information and insights into how IoT is impacting different industries, this knowledge will help you understand implications for your business.
2. This is your business and you alone are responsible for its success; partners play a part but cannot replace your thinking and efforts.
3. The partner and technology landscape is fast evolving, so you cannot rest because you made some choices when they seemed right; you have to be on constant vigil to see how best to serve your needs in this embryonic environment.

References

1. (Buyya and Dastjerdi, 2016) Rajkumar Buyya and Amir Vahid Dastjerdi "Internet of Things, Principles and Paradigms", Elsevier, 2016, ISBN: 978-0-12-805395-9.
2. (Deloitte, 2015) Deloitte, The Internet of Things Ecosystem, https://www2.deloitte.com/global/en/pages/technology-media-and-telecommunications/articles/internet-of-things-ecosystem.html
3. (Poter and Heppelmann, 2014) Michael E. Porter, James E. Heppelmann, "How Smart, Connected Products Are Transforming Competition", HBR Review, Nov. 2014.

Chapter 12
Developing and Improving User Experiences

> *Design is a funny word. Some people think design means how it looks. But of course, if you dig deeper, it's really how it works. The design of the Mac wasn't what it looked like, although that was part of it. Primarily, it was how it worked. To design something really well, you have to get it. You have to really grok [understand intuitively] what it's all about. It takes a passionate commitment to really thoroughly understand something, chews it up, not just quickly swallow it. Most people don't take the time to do that.*
>
> —Steve Jobs

12.1 Introduction

A holistic way to think about IoT is things, people, interactions, and outcomes. Often we get mired in the technology, but success requires us to look beyond the technology and applications [1]. Superlative success in building IoT-based businesses requires a strong focus on experiences and outcomes. IoT is significantly increasing the interaction users are having with devices and applications, often for much longer durations than ever before; this is driving user experience (UX) to gain new grounds. There are a few elements which are impacting UX more than ever—miniaturization of devices and displays, a deluge of data, shortening attention spans, interoperability between engagement platforms, increasing expectations of personalization, and the need for quick high impact. These factors create an interdependent tension in designing products as well as user experience. Everybody talks about how simplicity should drive design, but there is little guidance on how to bundle elegance, simplicity, and sophistication into designing user experiences, especially for IoT devices and applications. This chapter helps you with that.

We shall cover the following topics in this chapter:

1. Understanding how IoT drives design and user experience differently
2. Recognizing the psychology and science behind user expectations from product or service experiences

S.R. Sinha, Y. Park, *Building an Effective IoT Ecosystem for Your Business*,
DOI 10.1007/978-3-319-57391-5_12

3. Implementing converged customer experience
4. Leveraging user experience and design to drive differentiated growth
5. Building product design and digital design languages
6. Training your organization on design language
7. Maintaining your design language system

12.2 Understanding How IoT Drives Design and User Experience Differently

Since user experience is at the core of successful IoT businesses, let us understand the different influences impacting our design choices:

1. **Lowering attention spans**—In 2015, Microsoft did a study on the lowering human attention span. The study found that average human attention has reduced from 12 to 8 s in the preceding 15 years. This was caused primarily by the rapid growth of smartphones. 24 × 7 media, global interconnectivity, and the rise of social media also play an important role in this change. In today's fast-paced life, we have to multi-task. The study also found that our ability to multi-task has improved significantly, primarily buoyed by new technology advancements. This environment requires us to design which drives more focus and is able to maximize capturing user attention swiftly with high intensity.
2. **Harmony of experiences**—In nearly all cases, we can expect users to interact with multiple delivery methods while interacting with our IoT solutions and businesses. We access applications on devices, through mobile devices and sometimes through larger format computers. The applications could be run on multiple platforms and operating systems. Users expect each of these different interactions to lead to similar seamless experiences. We expect maps to look and work the same way whether we access through our vehicle navigation system or through phone or through a laptop. We expect reports and dashboards to be almost similar whether accessed through phones or desktops. Transporting experiences gets complicated because of the differences in the display and processing capabilities of the underlying hardware. Sometimes we distribute portions of experiences across different delivery platforms; we may limit the amount of data displayed or interactions allowed to compensate for the delivery platform limitations. Users may accept such limitations, but will still expect to have a seamless interoperability of distributed experiences.
3. **The experience of outcome and service**—In the pre-IoT world, we accepted differences between product experiences and service interactions; we stitched together a desired outcome through a combination of both. Today in the IoT dominated the world we have blurred those differences; we expect physical products, services built around them, and all interactions in-between to lead us to a desired outcome effectively and quickly. Not only do we expect the outcome to happen, we also want it validated and communicated in a very pro-

nounced way. This poses new challenges for user experience design because the number of variables and their combinations you have to deal with has amplified. Sometimes the expectations might drive conflicting requirements on the product or service infrastructure. For example, you do not want a call center to have a long list of automated questions; when you call, you want to speak to a real person quickly. But if the system does not first understand your needs, a lot of time will be spent on identifying the correct agent or you will need a lot of agents who are super-specialists. Your demands of much higher performance and capabilities from a device leads to cost escalations and sizing issues that may not be desirable either. You want everything your device knows about you to be reflected in all service interactions. An integrated experience of outcome and service necessitates user experience design to consider a much broader and longer horizon of lifecycle interactions that is very challenging.

4. **Personalization**—Our everyday lives have been complicated by too many things clamoring for our attention. Prioritization becomes a constant challenge and process. Also, the society has become more person-focused than commune oriented. In such a situation, we are always looking for 'what's in it for me' and 'what appeals to us more'. The issue is not linear because our preferences are neither static nor linear; they evolve over time and through new experiences. This makes it very difficult for manufacturers and services providers to first baseline then keep track of and then finally create unique experiences that meet the expectations of individuals, often on a real-time basis during delivery of product or service. In your design, you will need to balance such personalization with potential concerns around privacy and not becoming spooky or annoying. There are many technologies like embedded click analytics which have evolved to address such needs. This is one area where machine learning and deep learning has a lot of roles to play.
5. **Technical expertise**—Two groups of stakeholders involved in the design process pose different challenges due to rapid IoT growth—designers and users.

   Previously most users were expected to have a basic minimum technical expertise around using a particular product or service, be it of consumer or industrial type. Companies took guidance from technology adoption curves and relied on training efforts to get potential users acclimatized with their offerings. Rapid growth momentums of new IoT offerings means newer groups of users becoming consumers of those offerings, who often have very varied levels of technical sophistication. For examples wearables have target markets in rural under-developed areas as well in technically savvy urban youth population; companies need to build common platforms for meeting the expectations of both these markets. This poses huge challenges for design.

   Even for designers, previously they could rely more on art, intuition, and psychology. Now they need to equally account for insights from data sciences. Research is moving from more traditional feedback based methods to experiential analytical methods. IoT requires designers to master more disciplines than before.

6. **Deep analytics and recommendations to users**—Previously there used to be a set of specific interactions between the users and the product or service offering. Designers had to account for the most optimal workflows and best experiences.

   IoT is driving the need for more adaptive interactions between product/service offerings and users which depends on weaving in deep analytics and recommendations to users into the offering itself. This is thought-provoking because the offering does not have a static nature anymore, it has to become adaptive based on usage. So you need to account for a much larger set of possibilities during design around scenarios which might not have been encountered before.
7. **Design for rapid change vs design for longevity**—Traditionally design was expected to be timeless, now we expect the design to be timely based on changing usage and preferences. Iterative design processes came into vogue because of the rapidly changing technologies. This is how great design-led companies to keep up with change. This becomes exciting because you have to balance reuse, rapid development, investments, and returns. Extreme componentization in design is instrumental in such an environment, but it is difficult to simulate how the components maintain harmony and yet can be evolved over a longer period of time.
8. **Influence of wearables**—In the last 10 years, the massive proliferation of mobile devices made a huge impact on many aspects of our lives including design. In the past 10 years, there has been a 6000-fold increase in mobile data traffic. The introduction of smartphones 10 years back has completely transformed the mobile devices industry. Smartphone population globally has crossed two billion in this time. Today the smartphones we carry around have better computational capability than most commercially available computers from 15 years back. Mobile first has become a commonly practiced design paradigm in the last 6–7 years.

   IoT applications and services have been largely mobile oriented. Often the mobile devices become the primary method for user interface beyond limited interaction capabilities on the IoT device itself. For example, Fitbit shows a few summary and important data points and deeper analytics is presented on your mobile phone. In the industrial space, you see similar trends. Increasingly sensor and controller manufacturers are using mobile as the display and gateway device.

   Recently we have seen the increasing influence of wearables combined with this mobile trend. Wearables take the mobile experience up another notch. Apple watches, Motorola watches are prime examples of this. We are not sure that wearables provide any distinct advantage for users per se. This trend we believe is being driven by a perception of being 'cool' amongst consumers, and by category convergence being leveraged by manufacturers. In some sectors, wearables might offer some unique capabilities and benefits like in healthcare or buildings. We are still in the early days to understand the full scope and impact, but the trend is clear. This presents interesting design challenges for several reasons:

- Much less real estate for hardware and display
- New forms of interface
- Always on
- Potential interactions with human body

9. **Haptic response**—Haptic response centers on simulating experiences using vibration and similar waveform motions. Often tactile sensors are used in haptic devices to create/detect the simulation mechanically. Early explorations started around haptic technology more than 40 years back; momentum around applications has picked up only in the last few years. Currently, we see this technology being applied into gaming, VR/AR, computer hardware, robotics and mobile phone industries.

   Since haptic technology can be easily miniaturized and integrated with wearables, the rise of IoT-based wearables will drive increased usage. For example, haptic response embedded into a watch form device can be used to trick your body into a different state of temperature or other forms of experience.

   UX design for haptic response considerations presents a whole class of new challenges as you have to now deal with other types of sensory and feeling considerations that are difficult to define in abstract terms, measure consistently and concisely, and transport easily across different users. Users experience is usually in the context of variety of other physiological, psychological and environmental factors; haptic responses add another dimension to this list. The more we get into such domains, dealing with dynamically generated experience perceptions, and adapting products and services to user experiences becomes a huge challenge.
10. **Periscoping**—Periscoping is a very recent phenomenon, only a little over 2 years old. It practically started with the launch of the company Periscope in Jan 2015 and later acquired by Twitter. In our hyper social media world we are not only satisfied with live message updates, we want to get as close to the experience as possible through live streaming. Usually, such streaming happens from mobile devices and viewed over multiple format devices, mostly mobile, though.

    Periscoping has impactful implications for social networking. Greater connectivity between people and devices along with the rise of mobile devices will only increase possible applications of periscoping which is currently focused mostly around news and event sharing. For example, in future, we can foresee more real-time experience sharing and feedback using this type of technology. For example, technicians can "look-through-the-eyes" of users and guide them through troubleshooting or simple repairs without rolling a truck and making house calls. IoT technologies are simplifying communication and increasing affordability, so we see the clear growth of periscoping.

    This too presents new UX design challenges because your ability to contain or direct the experience of users is limited. Collective social preferences and mass-communication can very quickly impact your business. Whether you have periscoping features in your product in future or leverage the technology to get feedback or use it in any other format or are just impacted by it, you will have to consider such scenarios in your design considerations.

12.3 Recognizing the Psychology Behind User Expectations

User experience (UX) is defined as "a person's perceptions and responses that result from the use or anticipated use of a product, system or service" by ISO 9241-210, the international standard on ergonomics of human system interaction. So clearly user experience is largely about psychology. Since every user's psychology is different due different motivators and inhibitors, it makes UX very challenging because you have to solve for many possible scenarios and contradictions simultaneously. To win with superior user experiences, we must first get deeper into the psychology behind what drives preferences.

12.3.1 Drivers for User Expectations

Many scholars and practitioners have classified the drivers for user expectations and preferences differently. In our view, following are the four major drivers:

1. **Habit**—This is the effect of our past experiences and preferences. Over long periods of time, we get used to doing something or using something in a certain way. Consequently, we expect future products and services to work in a similar fashion. For example, the screen of every desktop, laptop or mobile device is usually rectangular; this emanates from thousands of years of reading from rectangular format printed materials. Habit causes behaviors to become subconscious causing implicit expectations. Habits train our minds to filter between good and bad designs or good and bad experiences. Habits have cultural and regional connotations. Good design or experiences for products in Europe have very different connotations compared to North America or Asia. While certain habits carry forward over generations, some of them are more impacted by demographic composition. A millennial will most likely use a certain product for much lesser time than Gen X making your design longevity requirement to be different. Habits can change but will take other drivers and time. For example, historically watches used be circular, but now you can find rectangular ones to be equally popular.
2. **Incentives**—Incentives are inducements for changing our habits. Such enticements can come in the form of rewards like money or products or services that we desire, or could also be in the form of simplifying some tasks for us. Nest thermostats induced customers to pay more for an unknown smart thermostat by leveraging a differentiated design, the much improved user experience of being able to control your home environment from anywhere in the world, and subsidies from utility companies. Similarly, smart watches are capturing people's imagination by combining multiple functions, simplifying them, offering better user experience, all at a reasonable price point. Sometimes rewards are membership oriented, i.e., using a product or service differentiates us and creates a sense of

accomplishment. Shopping at certain places, carrying some products or brands are examples of that. Another part of incentives is gamification—giving users badges and merits for a job well done. Fitbit will give a small fireworks display for people who take 10,000 steps in a day. Nest will display a green leaf badge if you are being energy efficient with your thermostat, etc.

3. **Guidance**—Guidance is targeted a campaign to change individuals' habits and thinking by using sublime marketing. In the industrial sector, RFID technologies, which in a way laid the foundation for future IoT technologies is a good example of changing inventory management and logistics management practices through guidance. The increasing popularity of Amazon Alexa is another example on similar lines; every day, every week, newer things are getting integrated to Alexa and consumers are now changing their habits from interacting with individual devices to a central general purpose interface.
4. **Hype**—When there is a mass hysteria around a particular product or service, we often get drawn to such phenomenon because of our desire for greater social conformance or not missing out on something. While guidance is targeted towards individuals in a push manner, hype creates a pull to which individuals want to get associated with. There are numerous examples of where hype has caused users to change their preferences, social media often plays a big role in this change. Every development around IoT in a way is a response to the hype around it. The popularity of periscoping is an example of hype influenced behavior change, instead of reading about events or watching news channels, now we are watching events real-time, curated by unorganized individuals or small groups of people.

   These drivers are not independent of each other, rather there is a lot of interdependence and cross-influence. This is an easy framework to remember; in order to get somebody to do something different, understand their HIGH! (Fig. 12.1).

   There are several other models to understand behavior and drive changes. Stanford faculty and researcher B.J. Fogg [2] has created the Fogg Behavior Model in which he describes that for any behavior change to happen, there must be a simultaneous convergence of motivation, ability, and trigger. Below is a depiction of his model. This is a useful tool to think about user psychology for influencing design (Fig. 12.2).

12.3.2 User Research

User research is critical to understand the preferences and expectations of users. This is the process to understand how design in its broadest definition will impact the user adoption of products or services. The psychology and science behind user research are quite evolved and getting better by the day. Christian Rohrer [3] created an excellent 3-dimensional framework to contextualize the different types of popular user research methods. His three dimensions are:

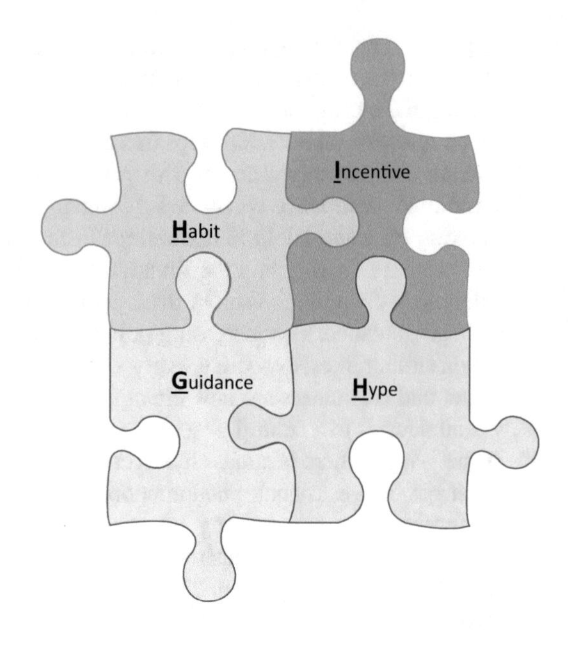

Fig. 12.1 The HIGH framework for influencing behaviors

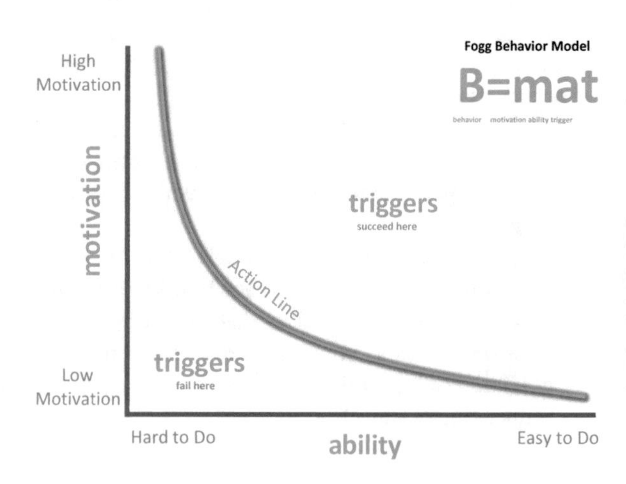

Fig. 12.2 Fogg behavior model

1. Attitudinal vs behavioral—what people think or believe vs what they do
2. Qualitative vs quantitative—observed directly vs derived information
3. Context of use—natural, guided, perceived, any combination of these or none

The following figure developed by Rohrer puts in perspective the various research methods against the above three dimensions:

A number of these methods evolved independently for interactive software and physical products. In the case of IoT, both converge and go beyond in integrating into lifecycle experiences of those products and services. So when designing the research programs, adapt the methods to account for your unique product/service scenario. In most cases, you will find a combination of methods to be most optimal for getting the best insight into users' minds.

If you are introducing a completely new product or service category for which there is not much historical precedence for usage or preferences, you may want to apply a method we call 'toddler tryout'. In most of the research methods in Fig. 12.3, there is some element of training and familiarity provided for the product or concept being researched. In toddler tryout, you provide no training or orientation to your research participants, you just give them a physical model or simulated environment of the product/service and see how they interact with it for an extended period of time, as if they are toddlers. You can design additional series of activities like asking the participants to reverse design the product, sell it, explain it to other toddlers; just do not ask for direct feedback on what they like and what they don't. This will help you understand habits, get clues about possible incentives and give you leads about how to design guidance.

User research is a very specialized activity. You need to have professional researchers conduct the exercises. Training people from other disciplines is always

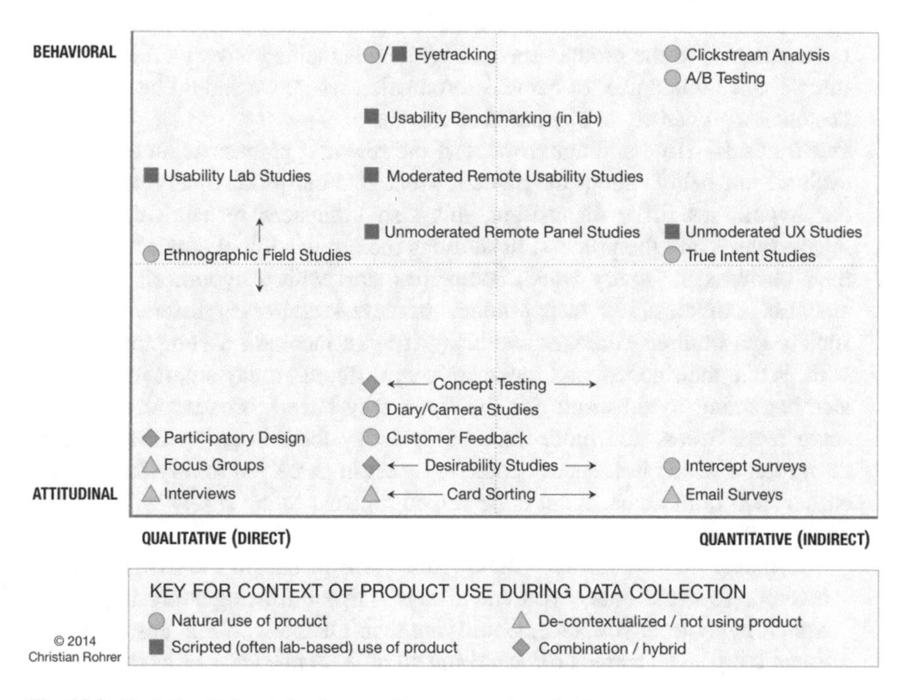

Fig. 12.3 Christian Rohrer's landscape of user research methods

an option, but sometimes sub-optimal. There are a lot of new tools to help you with research that takes away the human intensity and biases, be sure to look for them.

Any heuristic based research method is always subject to interpretation. So it is natural for design or interpretation biases to creep into the research process. We have seen three common biases which should be avoided during user research.

1. **Demographic**—Groups of people with common attributes around age, income, gender, region, race, religion, and other such factors have often demonstrated common habits because of similar background and psychological conditioning. They often have common channels for influencing behaviors or preferences through incentives or guidance. Getting the true representative diversity of intended target market at scale is hard, so often we use either fewer sample sizes or use proxies; this creates biases. We accentuate biases through mental models around these various demographic groups, which often leads to the incorrect design of the research itself. For example, usually we associate IoT products and services with younger demographic groups, but the appeal for older demographic groups in many cases (like remote smart speakers, thermostats, appliances, etc.) is much higher because older people may have limited mobility and different learning orientation.
2. **Disposition**—It is very common for us to be biased by precedence or how things have happened in the past. Disposition majorly influences our design of research and interpretation of results. For example, in trying to research users' preference for the name and features of a product, we generally ask for ranking. Our underlying assumption is that is customers like the name and features, they will be more attracted to the product/service. While this method is very useful for features, it has limited use for name determination where we should be evaluating customer's propensity to buy based on name alone.
3. **Destination**—Time and again we start the research process as an exercise to evaluate our beliefs about the market, users, and our product/service, and test our hypotheses. Like disposition, this also influences research design and injects biases into the process. In studying the implication of smart watches on their business, if luxury watch companies start with a hypothesis that their business is timeless like their products because it addresses different strata of society (an implicit demographic bias), it is an incorrect assumption to start with. While their appeal and looks are very different, today smart watches are also beginning to substitute for the style; they have a broader affordability; serve more needs, and more importantly today there is an upward mobility desire trend which influences people to invest in perceived luxury items much earlier than in the past. What is perceived timeless today surely will be challenged with time.

   Avoiding biases is difficult and requires diligence from the design stage of research. To avoid biases you can always start with listing every implicit or explicit hypothesis you have, identifying their alternatives, and ensuring your research process accounts for exploring all of them in an objective manner.

12.4 Implementing Converged Customer Experience

In today's hypercompetitive sonic speed world, customers have increasing choices for every need, keep evolving their needs constantly, and have decreasing attention span for companies to meet their needs. Companies are being challenged continuously to understand their customers better and serve them better. User experience has traditionally been defined as the user's interface with the product and service. But customers are now demanding excellence in the operational experience with the IoT products and services which have a little broader definition and often includes aspects like quality, accuracy, reliability, consistency, security, performance, speed, amongst other such attributes. They expect products and services to meet their requirements without exception every single time. The user interface—be it the screens in a software application or the human device interface in a physical product has often been the window into creating a lasting relationship with customers. The advent of industrial and intuitive design has served this goal well. More and more solutions have easier navigation, contemporary look, and feel, faster responsiveness, dynamic adaptability to create a strong impression on users. Everyday consumer influences like those from the mobile industry or social media are also impacting more focus on user experience excellence. But customers are demanding more today.

Customers are now also looking for excellence in operational experience. They require a high degree of quality and reliability. They expect the solutions around functionality, performance, speed, security, accuracy amongst others, to meet their requirements without any exception. If customers buy multiple solutions, they expect consistency in design and usage leading to a similar experience. Companies like Apple and Sony were the pioneers in solving for this, now Google and Microsoft are making good strides in this space. Customers also look for solutions to anticipate how they will use the solution and accordingly modify the behavior of the solution, and they want these behaviors to be optimized. These various dimensions—consistent product quality, optimized behaviors, consistent design principles and unified user experience contribute to the operational experience of the customer.

Along with our colleague Vineet Sinha we developed this concept of converged customer experience (CCx)which is achieved through a combination of user experience (UX) and operational experience (OpEx) summarized by the following equation:

$$CCx = UX^{\mathrm{OpEx}}$$

Since the value of the solution is primarily achieved by its usage and frequency of usage, OpEx has a multiplier effect for CCx. To achieve CCx, we need strong disciplines of customer behavioral segmentation, modeling interaction of customers with the solutions, playing various usage scenarios, competitive benchmarking, usage of data analytics, and finally keeping in mind the impact of other consumer experiences that influence customers. Many of the tools described in the previous section on user research are well suited for this purpose as well. Clickstream analytics, eye tracking, unmoderated UX studies, ethnographic field studies are good methods to understand the user's operational experience of a product or service.

Traditional technology approaches have focused on getting past experience static feedback, which always does not accurately point the company to the underlying experiences and expectations. Therefore traditional approaches have limitations in solving this problem. There are huge volumes and variety of data that needs to be collected, stored and analyzed to understand customer preferences and feedback. The feedback has to be constantly captured while customers use the solution—be it in the form of a web application or a machine or a process. Bits of these in-operation data by itself may not convey any significant meaning, but their aggregation or combination with other perspective data can convey a lot. On top of that, the context of the data can change over time with changes in customers' other expectations and behaviors. This means that we need to store the raw data of operational experience and user experience for a long time. The sources of data also can be structured, unstructured and streaming, something traditional technologies are not best adept at handling. Possibilities are endless and today we have the tools.

A lot of good research and framework development has already happened with respect to user experience design. In 2000, Jesse James Garrett published his model in 'Elements of User Experience Design'. In the 2015 book 'Designing Connected Products', Claire Rowland built on that model further. From that we would like to draw your attention to four intertwined components of user experience and operational experience design that are very relevant here:

1. **Interface design**—layout, look and feel primarily for interactive software
2. **Industrial design**—form factor for hardware products with or without display
3. **Interaction design**—architecture and behaviors of users interaction with product/service
4. **Interusability design**—interactions spanning across multiple platforms and devices with distributed capabilities

Because of the nature of products and services enabled by IoT, all of these four attributed have a symbiotic effect. Every choice in one has to be gauged against the implications in others. A simple and very common example of this interdependence implication is the responsive design of software capabilities. By definition, responsive design is seamlessly transportable across devices and platforms. However, if your UI and interactions have any tie-in to the underlying operating system of the design, you cannot truly transport it seamlessly. We have seen this happen several times with large industrial companies which have complex graphics and visualizations.

Earlier in this section, we had briefly touched upon designing for adaptability keeping in view converged customer experience. This is incredibly difficult because making the underlying infrastructure for any physical or software product adaptable is very expensive and nearly impractical considering any mass production or mass usage considerations. To achieve 100% adaptability, we will have to make every product or service a unique work of art that is individually refreshed frequently, like a custom made Ferrari which you a retrofit every year or so. The enabling technologies and economics are not favorable to such an environment, but customer expectations are moving towards that direction.

In the meantime, there is a lot you can do with extreme modularity in your design. The more your product design becomes a function of assembling reference products and components, the more you can adapt to individual customer preferences. This is true for software, hardware and hybrid products, even services. Platforming or core technologies that go into your product helps with this. For example, creating dashboards and reports required extensive programming 10 years back, some few years back and virtually none now if you use tools like PowerBI from Microsoft. Changeable faceplates and power modules for electronic products like phones, thermostats, other devices with displays are other examples of high modularity design. In many ways, this is a simple and quicker way to design. Every time you take this approach, you will have to pay extra attention to the four design aspects mentioned above. To be effective in such design, you will need a strong design language infrastructure, something we will get into later in this chapter.

12.5 Leveraging User Experience and Design to Drive Differentiated Growth

As we sift through history, we find many examples of products where their disruptive designs catapulted them to leadership positions driving differentiated growth. Apple under Steve Jobs was the beacon for such examples, but there are more. IoT has accelerated the pace of such disruptive products because now we can use technology and connectivity to solve customer problems differently. Let us review a few examples.

Nest disrupted not only how thermostats are perceived in the buildings industry, but also how energy management industry works. Nest inventors gave mobile control to people for managing their home environments. While the savings potential is impactful, it first caught people's attention because of a unique design and different user experience of environment control through smartphones. The ability to easily access and understand trends around energy consumption, integrate with demand response programs was distinctive. Nest created an emotional connection with the occupant which was never the case before; Nest made the boring thermostat cool. Now every building systems provider and energy management company has changed their approach for this class of devices and services.

Amazon Alexa has a similar disruptive impact on consumers, it created a unique UX without any UI. This was a paradigm shift for the industry. Now Amazon is making Alexa a major integration platform at homes and beyond for every kind of devices and services including searches for information. The product looks very simple and elegant, the user experience is a killer; to make those things true, a lot of thought and hard work was put into the design.

Another cool product we have been very impressed with is August Smart Lock [4]. Their branding goals of simplicity and security are achieved in a very disruptive way in an industry which has not fundamentally changed in over 100 years. Users do not have to worry about the hassles of managing multiple keys, fear to lose them if lost go through multiple hoops to get access to your door and get replacement keys, deal with insurance and so on. August took all of that complexity out by enabling users with digital keys which can be shared for limited use using smartphones. Now you can remotely engage with your security system and guests through your phone. Another paradigm shift achieved through an inimitable design and user experience.

Such examples are not limited only to consumer industries. There are a lot of very interesting examples of large industrial companies and even more traditional industries like agriculture. If you want to see disruption, just search about connected cows!

When we reviewed these designs carefully, we saw some common attributes that set them apart and most likely contributed to their popularity:

1. Iconic designs—people feel proud to own and display them, the product and its associated brand created an identity
2. Clean, classy, simple look—there is inbuilt elegance in each of them
3. Transported experiences—when experiences outside of the product had to interface with it, the flow was natural
4. Leveraged technology capability and perceptions to the maximum
5. Solved user problems in an impactful manner, was easy to use

So as you can see, good design matters!

12.6 Building Design Language System

12.6.1 Definition

Design language system (DLS) is an encapsulation of rules and guidelines to drive consistency and harmony in design across a group of products and services which could comprise of physical products, interactive software, and digital products that are a combination of both. Service experiences around such products are also covered by design language. DLS is not a collection of abstract concepts, is a practical and definitive guide for implementation of design ethos into actual products and services.

The purpose of design language extends beyond common design implementation. DLS also helps with:

1. Creating positive emotional connect with customers
2. Promoting brand identify
3. Differentiating from competitors
4. Reducing cycle time to develop new products by using standards and reusable artifacts
5. Onboarding designers and engineers quickly and uniformly

6. Allowing transport of experiences across products and platforms
7. Enabling your products to evolve over market, pricing and timing spectrums
8. Facilitating users to interact with your products and services more efficiently and enjoyably
9. Driving cost of product down
10. Integrating culture across internal organizations, channels, and customers

A good design language is characterized by many attributes, it is:

1. Aspirational
2. Motivating
3. Instructional
4. Comprehensive
5. Implementable
6. Binding

Design language efforts are deliberate and must have sponsorship from highest levels of the organizations. Design influence transcends multiple departments and functions in a very profound way, often requiring fundamental changes in many of them; which the executives can help navigate. Design language systems must intuitively communicate and establish the brand DNA of the organization and the product.

12.6.2 Components of Design Language Systems

There are at least three distinct layers of design language systems due to the inherently different nature of things they support:

1. **Physical product design language**—this includes any product with or without a display and extends to visible or otherwise aspects of products
2. **Interactive software design language**—this includes to any interactive software which exists independently or is instantiated through physical digital products
3. **Service experience design language**—this extends to every interaction customers have with your product support and enabling organizations including customer contact centers

There are many elements of design across these three different DLS layers, some of the more common ones are:

1. **Brand identity**—These are the human-like characteristics that can be associated with a brand. For example, Apple brand stands for simplicity, elegance, and intuitiveness. Apple's brand is also associated with differentiation. Almost to the point of creating a cult. If you review the evolution of its logo—the bitten McIntosh apple, it shows increasing simplicity and elegance. The attention to details in every aspect of the product and user experience including call centers forwards the perception of elegance and exclusivity. Similarly, Google's brand identity evolves

around global information accessibility, so every product from Google communicates that type of brand identify through easy, intuitive and fast searches.

2. **Design principles**—Design principles are a set of infallible guidelines that form the basis of designing products and services. These are higher level strategies that get implemented through rest of the design elements. Type ahead features in Google is a design principle that demonstrates their natural user experience and continuous learning philosophies. Creating magical experiences is a design principle at Bang and Olufsen which means the industrial design of their products has to create an instant lasting impression on the discerning aficionado.
3. **Visual design elements around color, shapes, typography, and iconography**—These manifest the brand identify and design principles in addition to simplifying the design process by eliminating any possible discovery or divergence around usage of the design elements. Visual design element guidelines will indicate the range of acceptable values to accommodate different sizes of product, display formats, and other environmental considerations.
4. **Information architecture**—This has two dimensions—how and where to collect, store and process information collected by products and services; and how and when to expose different types of information to users while experiencing the products and services. This is important because if you have different products and services, a uniform information architecture allows your users to have pleasant consistent experiences while reducing the need for training and even your internal development efforts.
5. **Training and documentation standards**—Apple advocates usage intuitiveness in product design, so it has eliminated training manuals and other forms of documentation. This is true for any physical digital product or software you buy from Apple. This also extends in a way to their call centers where you do not have to enter complicated menu options or be bounced around multiple people; a friendly and knowledgeable agent will understand what you were trying to accomplish, what issues you faced and walk you through the resolution.

Physical product design language will additionally include:

1. Guidelines for aesthetics, functionality, and ergonomics
2. Categorization scheme of different products into classes or families and sizes
3. Product artwork—industrial design and layout of different classes and sizes of products
4. Specifications for material, color, and finish
5. Assembly guidelines for products
6. Integration and interaction standards with other products including electrical, communication or other such interfaces
7. Interactive UI if applicable

Interactive software design language will include:

1. UI layout guidelines
2. UI components including symbol libraries
3. Wireframe designs

4. Content types
5. Workflow design
6. Animation
7. Audio interaction

Service experience design language will include:

1. **Initiation**—When the customer approaches with any service request either in person or by phone or through the web, how should the first contact experience be in terms of customer identification, time spent till contact, the information sought, etc. When you call a helpline and are required to enter a bunch of information, and then when you finally speak to an agent and are asked the same set of questions, it immediately worsens the service experience.
2. **Engagement**—When you interact with an agent or software system during your service experience, engagement defines what all formats and durations of experiences you should expect. This includes language, tone, and other auditory or visual cues that create an impression of the product and brand.
3. **Hand-off**—If one is transitioned between individuals or departments, those experiences are very defining moments for the customer. Service experience design guidelines for such hand-offs should identify all the guidelines to make such transitions pleasurable.
4. **Cross-selling efforts**—Whenever a customer comes in contact with an agent or interactive system during a service engagement, there is an implicit expectation to use that opportunity for trying to cross-sell additional products and services. While this is expected, sometimes it becomes annoying for the customer leading to detrimental results for the company unless clear guidelines are set around identification and approach.
5. **Feedback**—Taking feedback is important for any service engagement with customers. For any such feedback to be useful, its design must be simple, impactful and yet quick.
6. **Customer services metrics design**—Often we have seen customer service metrics stand in contradiction to the impression the company is trying to make through the products and services. For example, if the product is trying to generate a real-time feeling in usage, but every time you try to use a call center or online help if the time taken to resolve an issue or receive clarification is inordinately long, the brand identity is diluted.

12.6.3 Methodology

Developing design language is an involved activity. This has to be done in a very structured and methodical manner. We recommend using the following ten steps in developing your DLS:

1. **Objective alignment**—Multiple stakeholder groups in your organization will have different expectations from design language. You need to first establish a set of common objectives between these groups. This should include brand identity, design principles, layers and composition of design language system, the level of reusable artifact development and other major design choices. During this phase, you need to agree upon key stakeholders and engagement process with them.
2. **Team onboarding**—Once the objectives are clear, you need to identify and onboard the design language development team. As part of onboarding, you need a detailed association of the team with the objectives and provide context around guiding principles for making choices around those objectives.
3. **Portfolio assessment**—Next you need to gather insights around existing products and design guidelines. This should also include a detailed review of branding and style guidelines. The purpose of portfolio assessment is to get a better sense of what exists already and how it has evolved so far. The foundation for future is built on such an understanding.
4. **The user, market, and competitive research**—Now you need to get detailed insights into user needs and wants, understand how they interact with the products in their natural environment, find out how they feel about various design options, and ascertain their perceptions around what is good and desirable. You must also evaluate how the market is evolving, not only for your industry but also adjacent and other influencing industries. For example, if your business has something to do with communicating devices, it is useful to understand the trends and implications from smartphone industry. Gaining insights into competitors' actions are is critical to design your defense responses or attack responses. For example, if your competitors are significantly ahead in mobile applications and you do not see a pathway to compete for their effectively, you may choose to migrate to wearables right away and lead the market there.
5. **Design strategy formulation**—This is where you formulate how you distinctly differentiate through design to win decisively. The strategy should include your long term plans for brand expansion. Design strategy should be an extension of your business strategy. A good design strategy helps you with risk mitigation for changing market conditions and evolving user preferences. It should also help you prioritize the preferences you are trying to create in your customers for your products and services. The output from this phase becomes the soul and guidepost for every other design element.
6. **Visual brand language (VBL) definition**—Now you translate the design strategy into discrete design elements like colors, shapes, iconography, typography and layout composition. At this stage, you will build the industrial design patterns, UI patterns, symbol libraries, and graphics to further your brand and design objectives. VBL helps identify and influence all the product attributes that impact your brand.
7. **Design refinement**—In this stage you refine the designs into actual instantiations of products and offerings. For example, if you are designing a class of

physical digital products, this is where you create the details taking into account your design strategy and VBL definition. During design refinement, multiple acceptable options are created for different types of design implementation.

8. **Tools development**—In this phase, you engage in automation of the design language system. The benefits of a design language can be amplified if there are reusable design elements like code snippets for interactive software or CAD files for physical products or interaction platforms for digital products. This leads to true cycle time reduction as you are eliminating work. At this stage, you should also explore developing tools for validating compliance with your design language and generating recommendations for fixes where there is non-conformance.
9. **Stakeholder validation**—This is the iterative process of seeking and implementing feedback from key stakeholders in the process. For effective validation, you will have to educate the stakeholders about your research findings, implications, and resulting design choices. You may often receive biased inputs based on design legacies, your portfolio assessment should help you sift through such biases.
10. **Deployment planning**—This is the last stage of your DLS development where you need to complete planning for implementation, training, marketing, and maintenance of your DLS. Here you design the success of your DLS initiative into practical implementations over a sustained period of time.

Developing a good DLS takes time, often several months depending on the complexity and expansiveness of your products and portfolio. This has to be a collaborative activity between various groups involved with product design and customer success.

Like user research, design language development is also a very specialized activity. If you do not have the right resources internally to do this, it is better to secure design language development expertise from outside so that you can do this fast and effectively. This way you will also be able to benefit from other experiences such expertise may bring from their past engagements. Usually, design principles and best practices are not limited to a particular industry and can be carried over multiple industries; external resources can help with that too.

12.6.4 Socialization and Review Process

If you are trying to bring substantial changes to how design is perceived and practiced in the organization, your socialization and review process for the design language system needs to be intensified compared to normal review processes. Here are a few things we recommend you should do:

1. Keep the executive layers updated and involved more than usual, you can use their influence to drive your agenda more effectively. If you can find a senior leader to champion DLS, your cycle time to successful adoption will be quicker.

2. Drive greater consistency in communication across organizational layers as it helps with unified changed management. Yes, continue to do abstractions for different layers of communication, but be consistent in your content and intent.
3. Make the communication as visual as possible, use props and models to demonstrate progress, after all, design has a lot to do with the visual appeal.
4. Relate design language development effort to other active product development initiatives and demonstrate the value of DLS for those product efforts.
5. Listen more, defend less. Whenever we are trying to show something new and we get feedback which is not exactly aligned to the output we have created, we quickly get into defending our process. Design changes things more fundamentally than other influences, so take cognizance of other people's feedback; they may have something really useful to offer, if nothing else, they will adopt more if you listen better.

Socialization and review process must be iterative to be effective. There is a fine balance between involving too many people too early because it can become very disruptive. So identify the key influencers and involve them. Typically your product managers and key commercial organization representatives like high performing sales people will be such influencers. You need a small representation of designers and engineers to keep things practical, but you will get sufficient opportunities to socialize with them during implementation.

Socialization should extend beyond your internal organization. DLS is an asset which you can employ for external marketing and branding initiatives.

12.6.5 IP Review for Design Language

In every aspect of development related to IoT businesses, there are numerous opportunities for intellectual property (IP) generation. IP helps you in many ways:

1. Demonstrate thought leadership because patents are globally acknowledged as a proof point for your forward thinking in technology
2. Protect your business against infringement by competitors, even if your competitors scale quicker, you can go after them
3. Creating higher barriers for success for competitors as they will now have to design around your IP

Generally, we think of IP when we are building computational algorithms, or developing software routines or are considering an innovative actual product design. Design patents are also important and can give you a competitive edge, but often neglected. In 2015, the US Patent and Trademark Office received over 37,000 design patent applications and awarded little less than 30,000 patents from previously received applications. This number has been rising due to a greater appreciation of design on hand and overlapping products on the other. To get a design patent, you must have a unique embodiment of an ornamental design in a manufactured product which could be a physical product or user interface. Software UI is usually harder to protect than physical products, but if there is true uniqueness, it is protectable.

Given how much of an influence DLS will have on your products and their design, it is important to do an IP review during the development of your DLS. Following should be considerations during IP review to evaluate if your designs will infringe somebody else's:

1. Design patterns and layouts
2. UI organization
3. Symbols, font, iconography
4. Design concepts
5. Interaction design

Post the IP review, you should consider the following for creating new patents for yourself:

1. New design patterns, symbols or iconography
2. Innovative design concepts like replaceable covers
3. Unique interactions and workflows

If you are a start-up company or larger business looking for external investments in future, remember that your IP portfolio helps with your valuation; it has tangible, accountable value that your investors care for.

12.7 Training Your Organization on Design Language

Any design language system is inversely proportional to the legacy of design in your organization. So simply put:

Design language adoption α 1/(design legacy)

Design legacy is contributed by many factors, notably:

1. Large portfolio of existing products
2. Longevity of your products
3. Historical organic diversity in design
4. Deep rooted belief systems and background of designers
5. Brand diversity
6. Tenure of your designers with a long legacy of design

Design legacy makes it hard for companies to change. Consequently, when we review examples of disruptive design led products, we do not see too many examples or companies with long legacies. This problem is acuter for large industrial companies. Given that to win in IoT businesses, you need a different approach to design and implement a new design language, in a legacy riddled environment, training your organization on design language becomes very important. Even when you do not have a legacy, but are racing against time and competition to win, getting a unified understanding of design language helps speed things up.

Training teams on design language require reinforced approaches in addition to normal training methods. There are several reasons for it, some of the prominent ones are:

1. Design language can be construed to hinder engineering creativity. Loss of independence in thought and implementation raises artificial barriers amongst engineers in efficiently solving a problem.
2. Proper internalization of design language requires an understanding of a broader set of abstract concepts around brand identity and reinforcing visuals, user behaviors and attitudes, subliminal influences of design on user preferences, and experience lifecycle management.
3. Application of design language has to be consistent throughout the business and engagement lifecycle, so you need to understand every aspect of the business.
4. There is more discipline required for implementing design consistently.
5. Design language will evolve, so you will need more adaptability in new improvements as soon as people start getting comfortable with the established design language.

To augment traditional methods like a classroom and hands-on training, you can employ hackathon type of formats. Design-a-thon events have started becoming popular over the last couple of years. You can invite designers, engineers, user research experts, customer service professionals, commercial organization representatives, product and program managers and representatives from other functional organizations into a common session where you expose them to your general design principles and guidelines, give them access to your entire design language repository and toolsets, and ask them to engage in some collaborative design workouts. These workouts could be for new products or make major modifications to existing ones. Same format applies for service offerings too. In this type of exercise you will be able to:

1. Promote cross-discipline collaboration without seemingly artificial interventions
2. Let people internalize the design language system through practical application without onerous learning exercises
3. Make design fun, interactive and iterative for people who may not have deep appreciation for the subject or much formal training in it

Another activity that Google has made really popular right now is a "Design Sprint." Design sprints are an intense 1-week activity for any project or product team to kick-start design on a project and give it an opportunity to be way out in front of the development. "*Sprint: How to Solve Big Problems and Test New Ideas in Just Five Days*" by Jake Knapp is a fantastic book on the subject.

It is very important that your executive layer is visibly seen in support of common design language socialization and implementation efforts. This helps send a message to the organization that design is critical to winning in the market.

Training your organization on design language is not a one-time activity, you should plan refresher courses frequently. You should also get external design experts and customers to come and engage with you on these refreshers—it will increase exposure for your teams, involve customers for better implementation, and create more buzz around your design efforts.

12.8 Maintaining Your Design Language System

Design language systems are like living systems that evolve over time and may decline without proper upkeep. Keeping DLS fresh and relevant is equally important as its development. There are several critical influences for periodic review and proper maintenance of DLS:

Brand identify refresh based on evolving corporate strategy, market perceptions, and business aspirations

1. Technology changes enabling new capabilities in products and services to deliver newer forms of customer value or technology obsolescence
2. Market movements
3. Changing user preferences around various design elements
4. Other industry influences
5. New hype cycles
6. Geo-political events impacting businesses and products

You should have a periodic cadence to review these influences and update your DLS based on what is relevant and impactful for your businesses. As an example, user preferences around person to person interaction might be tilted towards offline textual communication but interaction with devices and systems could be biased towards natural language enabled voice interactions. This creates different design implications for different types of interactions which may not have been the same during original instantiation of the products under design review.

12.9 Summary

As mentioned in the opening quote, designing user experiences is complex and needs to be comprehensive. In this chapter, we have exposed you to the various dimensions of design. In summary,

1. IoT brings a lot of new capabilities but equally intense design challenges
2. Factors which influence design for IoT products are rapidly evolving and need to be carefully evaluated for incorporation
3. Psychology has a key role to play in determining user preferences and expectations from products and services

4. Customers are looking beyond user experience, they want a converged customer experience which includes operational experience from products and services
5. Differentiated design and user experience can be leveraged to drive exponential growth
6. It is best to capture design principles and approach through a comprehensive design language system which includes reusable product development components
7. Design can be an important contributor for your IP assets
8. Building a design language system is not enough, you need to train your teams on it and you need to actively maintain it

Designing Connected Products by Claire Rowland, Elizabeth Goodman, Martin Charlier, Ann Light and Alfred Lui is an extremely good book you could refer for more implementation tools and techniques around designing for IoT-based products.

Happy designing, now let us move on to marketing your IoT business and products; if done right design will help there as well.

References

1. (Allen) Allen, J. J., & Chudley, J. J. (2012). *Smashing UX design: Foundations for designing online user experiences* (Vol. 34). John Wiley & Sons.
2. (Fogg) "What Causes Behavior Change?" Behavior Model by BJ Fogg http://www.behaviormodel.org/
3. (Rohrer) "When to Use Which User-Experience Research Methods" by Christian Rohrer https://www.nngroup.com/articles/which-ux-research-methods/
4. (Scalisi 2015) Scalisi, J. F. (2015). U.S. Patent No. D727,769. Washington, DC: U.S. Patent and Trademark Office.

Chapter 13
Marketing Your IoT Initiatives

> *Here's to the crazy ones, the misfits, the rebels, the troublemakers, the round pegs in the square holes… the ones who see things differently – they're not fond of rules… You can quote them, disagree with them, glorify or vilify them, but the only thing you can't do is ignore them because they change things… they push the human race forward, and while some may see them as the crazy ones, we see genius, because the ones who are crazy enough to think that they can change the world, are the ones who do.*
>
> —Steve Jobs, Think Different (1997)

Businesses and customers view the benefits of IoT differently. Everybody has a bit unique perspective on this. It is critical that you market your IoT efforts effectively to your internal stakeholders and your customers. This is important because you need to educate stakeholders about the inherent change management issues in how you transact business in the IoT context. You also want to demonstrate your thought leadership and technical stewardship in bringing new value to everybody internally and externally. Like any other marketing effort, you need to exploit multiple channels of marketing and also embed marketing efforts in your products and services. In this chapter, we will introduce you a framework of driving thought leadership through a product-patent-publicity framework and expose you to some of the best practices in the field.

We shall cover the following topics in this chapter:

1. Understanding the significance of marketing for IoT initiatives
2. Introducing the product-patent-publicity framework to aid in designing your marketing campaigns

S.R. Sinha, Y. Park, *Building an Effective IoT Ecosystem for Your Business*,
DOI 10.1007/978-3-319-57391-5_13

3. Integrating IoT marketing efforts with your overall organizational marketing program
4. Integrating marketing into your product strategy and design
5. Managing social media marketing effectively

13.1 Understanding the Significance of Marketing for IoT Initiatives

IoT redefines the engagement between customers and products/services. Marketing solutions provider Marketo has an interesting perspective on the changing landscape, they believe "the interconnectivity of our digital devices that provides endless opportunities for brands to listen and respond to the needs of their customers—with the right message, at the right time, on the right device" [1]. Now data and analytics driven targeted marketing are becoming the vogue. Customers are bombarded with multiple marketing initiatives every time they are connected to the internet through any platform or product. It has become quite challenging for marketers, for the same category, customers often see multiple products/offerings with slightly different capabilities and messaging. Connecting with customer's real problems and being able to solve them with impact is key to success in the IoT world.

There are a number of elements which are different when marketing for IoT products and solutions:

1. **You have to often sell the idea first**—As IoT technologies enable new capabilities in products and services that people are not used to, often you will have to first sell an idea. This necessitates creative concept marketing. Let us take an example. Historically, in the last 30–40 years, the basic design of a computer and its peripherals have not changed dramatically. The form factors have become more stylish, the devices have become miniaturized, storage and processing technology has become compact and higher powered, but the fundamental design and functioning have remained the same. Recently a new computer called Solu was launched that has a very different design paradigm. It has one of the smallest form factors with no display or keyboard and a cloud linked operating system. The whole user experience is also very unique mimicking human brain functioning. You can buy the device, operating system, storage, and apps for a monthly fee instead of one-time device purchase and annual license fees. Consumers may not implicitly trust such a proposition because we have been wired to think that more powerful computers with better features and apps from reputed manufacturers are better. To counterbalance these mental models, Solu is messaging around social computing, transportability of computing capabilities and highly intuitive experiences; they are encouraging people to reimagine computers. In the industrial space, this is a more complicated issue because the pace of behavior and experience change in using technology is very slow. Experienced technicians and discerning customers initially find it hard to believe how machine learning can substitute for decades of human experience. So you need to be able to articulate value in a very different way.

2. **The product is a channel for marketing**—Historically marketing channels [2] evolved from print and public display to television to online and social media. The product was never really used as a channel for marketing because there were limited opportunities for connected interaction between the user and manufacturer via the product. IoT has changed that paradigm. In fact, the product is one of your best ways to market and message because the user will be focused on the product's context and not be diverted by noise which might be present in the other channels.

   This shift requires marketers to think of a perpetual engagement instead of campaigns.
3. **Adaptive marketing**—The product keeps morphing and how people use the products and services also might be different. Moreover, different users may be different versions of the product so their scope needs to be taken into account. In many cases, the definition of an IoT product is not static, so you may need to adjust your marketing efforts accordingly. For example, let's say you have a smart thermostat at home. This device has a service which allows utility companies to take control of your temperature set point within previously agreed limits to optimize energy consumption. If this service is an optional one, it should be marketed only to users who do not have the service and in areas where the utility companies have that choice. In such a dynamic environment, you need to adapt your messaging threads based on the customer and product. The success of marketing is therefore defined by the confluence of three things—technology (to help with specific targeting), creativity (to help with impactful targeted messaging), and discipline (to drive high top of mind recall).
4. **Data-driven marketing**—IoT is driving data sciences into marketing. You have to develop journey maps starting from every possible opportunity the customer may first come across your product, to all the influencing factors for her buying preferences, to the actual sale process, and further extending to ongoing usage and service. You have to analyze customers' behaviors and drivers for those influences on a real-time basis, and take marketing decisions at split second intervals on a very high volume of transactions. You have to capture insight and use them in context. You have to deliver the right message at the right time using the right device or channel. You have to manage to message but simultaneously think of experiences. These are entirely new ways of thinking about marketing. In the next 5 years, more than half of marketing efforts will be influenced by IoT.
5. **Social media impact**—Social media has a disproportionate influence of all marketing channels. It is always there and nearly everybody is on it. There are already too many social media outlets and more are coming. The demographic alignment of the various outlet is rapidly changing, while Facebook was very popular with millennials till a few years back, they are now shifting to Snapchat and Instagram. You can never control what perceptions get formed or fanned. Something might become popular instantaneously causing revolutions, others might get lost in the noise. Social media companies are constantly evolving their monetization and marketing models, so you have to always keep in sync. While it is a great channel for marketing, managing social media is a complex and constant endeavor. We will go into some best practices later in the chapter.

6. **Augmenting your IoT play**—Your business can have several identities for an IoT play—in a very simplistic model you can be a hardware manufacturer or a software provider or a services provider or have a play across more than one aspect. You will, however, have one dominant focus around which others pivot. Similarly, you will have a positional play of being a leader, challenger, or just another player still trying to find an identity. You growth strategies might focus on a particular demographic or geography. All of these different considerations define your IoT play and your marketing efforts have to augment them. This is critical for marketing program design because the methods employed to address them can be quite different from each other.

There are a number of new technology developments which also will transform marketing in an IoT dominated world:

1. **Alternate reality technologies**—As marketing transitions from more message based to experience oriented, alternate reality technologies can help influence those experiences. There has been a lot of progress in both virtual reality (VR) and augmented reality (AR) technologies making them more impactful, easier to use and affordable. We see a lot of new applications across both consumer IoT and the industrial internet where VR and AR technologies are employed—be it virtual room design or immersive product experience.
2. **Massively distributed and miniaturized sensing and compute capabilities**—In previous chapters we have touched on the technology advancements and new capabilities. Now we can put sensing and computing capabilities in almost every kind of device. This allows the device to learn the user's behaviors locally, which is great to distribute targeted marketing, very personalized experiences, and instant feedback.
3. **Big data technologies**—Previously we had to create metadata around customers and products across different engagement platforms or data sources. Now it is possible to synthesize data from multiple sources in different formats to create a converged view of the customer. This allows marketing campaigns to be more customer-centric than channel focused.

13.2 Introducing the Product-Patent-Publicity Framework

In the previous section, we discussed how some of the marketing paradigms are different for IoT. Over time, businesses like yours will become very proficient in exploiting the various channels and applying technology optimally, to get better insights into customers, create better awareness of your offerings, and influence buying behaviors. In this section, we would like to introduce you to an additional tool we fondly call as the 3P Framework for IoT. This can be a useful aid in augmenting your marketing efforts. But before we go into the framework, let us discuss its context.

In any business, to be successful, you need leading products and channels to distribute those products. To lead with products you need superior capabilities than others, differentiated technology and other attributes necessary for any successful product/service business. All of this is very well known and developed to the point of being a near science. Since the IoT space is very hyper competitive and evolving at hypersonic speed, too many people are constantly clamoring for the same user's attention. So your differentiation strategies cannot solely rely on features and capabilities, they have to be founded also on thought leadership and future orientation. If you are seen as moving the market, your chances of being the default choice are very high. For example, one can argue that today there are a number of smart thermostats in the market which are equally or more capable and impactful as Nest, but it is still the first name taken in any discussion around smart thermostats. As the entry barriers for new players in most product categories are lowered by the development of IoT technologies, now you not only need to be better than other alternatives, you also need to explore blocking out alternatives from the market or make it very difficult for them to develop and sell products.

This is where our product-patent-publicity or 3P framework is helpful. Think of the 3P Framework as a very well-oiled array of complementary contributors that help you become more visible and win (Fig. 13.1).

13.2.1 Products

Most companies have a strong focus on products—be it physical or service products. Product management is a very well established discipline. IoT brings some new dimensions to product management:

1. There are more technology interfaces and layers for the customer-product-company engagement
2. Engagement and experience take more precedence than the raw product characteristics

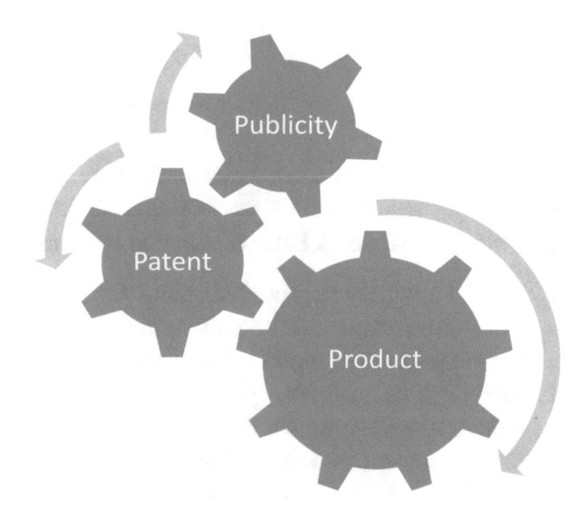

Fig. 13.1 3P Framework

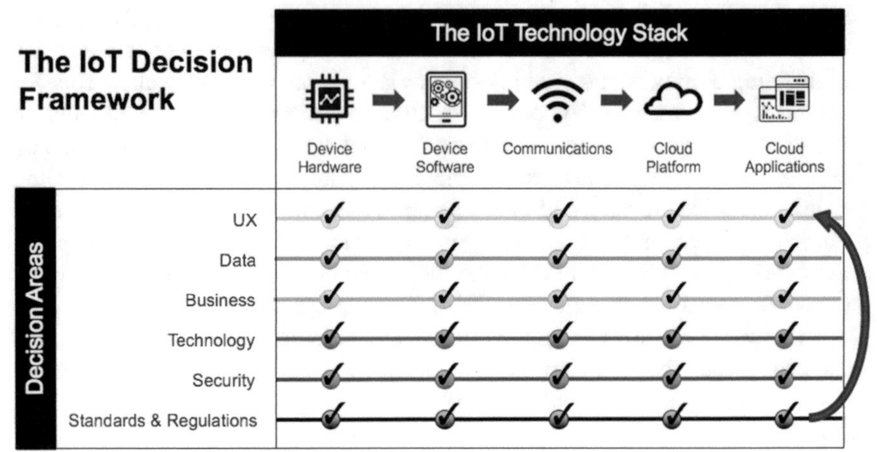

Fig. 13.2 The IoT decision framework

3. Multiple complex domain expertise is required to deliver a successful product
4. Keeping up with rapidly evolving customer expectations
5. Business models are more utility or consumption pricing oriented, requiring newer models to think through traditional businesses cases

The IoT Decision Framework developed by Daniel Elizalde [3] is a good tool to think about product management for IoT businesses. It lists the key decision areas across key components of the technology stack (Fig. 13.2).

We will not go into details of the dimensions or technology here as they are covered in other chapters. Given the specific context of your industry and markets, you need to identify probing questions against these categories to help you uncover customer value proposition and your differentiation strategy. These become key inputs to your marketing strategies.

13.2.2 Patents

Most companies get well focused on their product strategies and create brilliant marketing plans around those. Very few, however, harness the power of patents as a core business strategy. Patents help you in many ways:

1. This is a widely accepted standard for innovation and thought leadership. IoT and innovation are quite synonymous. A higher number of patents help your company get identified as a very innovative one.
2. The rigor of patenting process requires you to think very deeply about the uniqueness of your customer value proposition and solution. This way you will focus more on design, analytical and algorithmic aspects of the products and offerings.

3. Patenting is a great attraction and motivator for people with a research orientation. You need to attract such people because you may be attempting to solve problems in a new way instead of incrementally improving an existing process.
4. Patents create a huge barrier for your competitors. Either they will not be able to pursue the market in the same way you are doing or will have to create complicated workarounds that will impact their competitiveness.
5. If you ever get sued by a competitor or patent troll, the more patents you have helps with negotiations. Remember the famous Apple-Samsung IP battles?
6. Patents help your business attract greater financial value from investors as patents are believed to have inherent tangible value.

Every time you engage at the ideation stage of a new product or offering development, explore the patenting opportunities. You should have a portfolio map of patents (secured, processed, intended) for your business and review it few times a year involving executive business leadership, product management, marketing, legal, research and product development. You need to identify areas which provide you with a strategic advantage, then make them focus areas for future patent development. You should study the patent profile of your competitors and use those inferences to influence your business and patent strategies. Sometimes you will need to build defense postures to deal with potential competitive actions. We recommend that you should even have very specific numerical goals for innovation disclosures and patent filing as part of your performance management system. The intellectual property of your business demonstrated through your patent portfolio should be a key element of your marketing strategy.

13.2.3 Publicity

Your marketing department or external marketers must be well versed with the different publicity channels, methods and programs; so we will not rehash those here. What we would like to draw your attention is to become active in the various conferences, online forums, publications and other such channels. These are good tools for many reasons:

1. Showcase your personal thought leadership or those of other employees in your organization
2. Connect with other individuals in similar space and learn from them
3. Learn about innovative use cases and applications of IoT technologies into new areas
4. Expand your network of ecosystem partners
5. Get refreshed with a creative and additive outlet

Getting yourself and your teams out there (provided that does not start becoming a distraction from regular real work), helps you get the message out. Hey you are reading this book and possibly wondering about applications we have worked on!

Since IoT is such a big hype, there are numerous opportunities for publicity. Be thoughtful in choosing the forums which have maximum impact for either your business either in connecting with customers or seeking our suppliers or partners. You may not always find an IoT event or forum for your specific industry, in such cases, get associated with some of the generic ones.

13.3 Integrating IoT Marketing Efforts with Your Overall Organizational Marketing Program

If your business is a very diverse one with both IoT and non-IoT offerings, it is possible that there is some divergence in the marketing efforts between them. Such divergence may happen because of several reasons:

1. The IoT business is organized as self-contained and sheltered from rest of the organization to allow it to grow independently
2. You need slightly different marketing skills and orientation for IoT-based products
3. Your traditional part of the business has very well defined and established channels and corresponding marketing programs; your IoT business is in a more rapid evolutionary phase

If there is the possibility of such a divergence, you must act to bring them together. This is critical for many reasons:

1. If you have same products with and without IoT-enabled capabilities, one could cannibalize the other. If you have a planned migration of focus, your marketing efforts must be in sync to support that. Alternatively, you could keep pursuing both business lines at the detriment to each other.
2. Your brand identity might get confusing for the markets and customers. For IoT products, if you focus too much on rapid adaptability while you keep talking about near perpetual longevity for non-IoT versions of those, you can create confusion if the messages are not crafted well to reinforce each other by highlighting different aspects of the product.
3. Avoid wasted efforts, especially on common channels like social media. We have come across very few companies which claim to have enough people and money in their marketing budgets, if you are not one of them, you need to optimize resources for maximum impact.

There are many ways to solve this integration issue—organizationally, financially or by aligning agendas. How you do it entirely depends on the context and maturity of your business and marketing organization. It is important that at the outset you lay out the long-term plan and make that as the basis of any alignment.

Let us also not underestimate the amount of internal marketing that is required for the success of initiatives. This is a new space and you need larger organizational horsepower to support your efforts. We have dedicated an entire Chap. 8 in our book '*Making Big Data Work for Your Business*'. We recommend a review of that if you this internal and stakeholder communications is of interest for you.

13.4 Integrating Marketing into Your Product Strategy and Design

Marketing has evolved keeping pace with the changing nature of trade and commerce. Prior to the industrial age, there was very little need for marketing as most of the commerce was based on the trade of goods and services between producing regions and consuming ones. In the industrial, once more goods started getting produced the markets started taking a more global shape, marketing showed up on the horizon to assist with demand fulfillment. Over the last several decades as supply has outpaced demand, marketing has taken a more active role in demand creation and evolved very well as a discipline. IoT has exponentially opened up opportunities for businesses as well as customers, thus increasing the role marketing plays in the success of a business.

Traditionally marketing experience used to draw customers and drive demand. In the IoT world, marketing, and product experience have started to converge. Rapid changes in customer expectations, technology cycles, and product characteristics is further fueling this convergence. The design is emerging as the unifying factor for this convergence. In the previous chapter we have discussed design and user experience extensively, so here we will focus more on marketing.

Integrating marketing with your product strategy requires consideration over many dimensions, let us review some of the key ones.

1. Identify the core elements of your business growth strategy and review how the product strategy implements them; then design your marketing efforts to highlight these. For example, your business strategy rests on driving cross-sell and up-sell of subscription based services to an increasing captive customer base, and your product strategy implements this by offering new analytical capabilities every 3 months on your existing products accessible through the cloud and mobile platforms. In such a case, your marketing efforts must highlight the value of these new offerings in the specific context of the customer, such as how much more productive they can become or how much more money they can save, etc. by using these new analytics. The messaging must be delivered through the customer's choice of engagement platform and also through the product if there is an option for such distribution. The timing of the campaign must be synced with not only new offering launch but also with any major events the customer engagement cycle might have like budgeting or review cycles. Of course, there must be easy ways for the customer to buy these if she so chooses. In case the customer ignores the new offerings you are trying hard to sell, you should have some subtle messaging on what they might be missing out (again with tangible personalized value to the extent possible). These types of initiatives have to become part of your operating rhythm and not remain as isolated marketing interventions.
2. It is critical to highlight the elements of differentiation in product design during marketing and product usage. If attributes like simplicity and elegance are key elements of your product design, you should highlight those in the user experience and also when you talk about the product in marketing efforts. We talked

about Alexa in the last chapter. It changed the user experience paradigm by taking out visual UI and replacing it with natural language processing. The product is promoting an extremely user-friendly and user-oriented persona. So if you look at the design, it is simple yet inviting. When you see the marketing ads, you will find them to exemplify simplification of everyday common tasks performed by people like you and me. The technology behind Alexa or other similar ones is very sophisticated and advanced, but marketing hides all that complexity.

3. If you use a smartphone or other similar smart devices, you must be familiar with the pestering software updates. Most IoT offerings have periodic software upgrades to address bug fixing, security patches, and other such housekeeping activities. Till you do the updates, the system keeps reminding you; when you get around to doing it, you have to read through pages of legalese and accept terms you hardly understand. Most people view these as boring and inconveniences. But you can change that. You should use these events as a marketing tool to promote how you are enabling new capabilities for the customer and create another opportunity for engagement.
4. Earlier we could have marketers and product people specializing in their own domains; given all the convergence it has now become critical that marketers, product managers, and engineers have a very deep appreciation of each other's domains. Marketers need to have some reasonable understanding of the technology and product roadmaps to effectively craft messages and campaigns; product managers need to understand impactful marketing messaging to refine their product value proposition; engineers need to understand intended messaging so that those can be highlighted in the product itself. You need to promote more collaborative working between marketing and the other functions.

The cool technical capabilities enabled by IoT are captivating. Many IoT businesses get much focused on the technology initially, we have seen too many instances where such businesses fail to take off because they lack proper marketing involvement and clear customer value articulation.

13.5 Managing Social Media Marketing Effectively

Social media marketing is key to success for any upcoming IoT business. On surface social media seems easy, how difficult really posting things on Facebook and other sites can be? But it is far more complicated and needs meticulous planning. Here are some best practices that might be helpful for you:

1. **Strategy**—Strategy is very important because it defines the boundaries for your activities and sets expectations. Assume you have developed a new predictive diagnostic tool for use by expensive manufacturing equipment in the hi-tech industry. Your solution includes new sensors, data collectors, computational platforms, cloud analytics and mobile apps. It is very unlikely that you will find a lot of customers for this solution on Facebook. But you still need to market the benefits of this through Facebook as part of your overall organizational social

media plan. For building a long term strategy to drive impactful social media marketing you need to consider many elements like:

- Who is your target audience?
- What is the purpose of your social media marketing—lead generation, branding, awareness, tapping into customer psyche, etc.?
- What kind of impression do you want to create on your audience, what should be your tone?
- What kind of messaging do you want to focus on social media channels and if you want to use them for any exclusive pieces?
- What kind of momentum do you want to have in your social media marketing efforts? By momentum, we mean frequency of engagement on social media platforms. This has to be in line with your goals and also with your capacity to manage social media.
- How do you want to make your presence felt and enlist followers?
- How much money and resources can you spend on social media marketing efforts?

2. **Presence**—You need to be deliberate about which social media platforms you put your business on and what your intent with that channel is. Each channel has a different intent and design. In the example from above, you may choose to use LinkedIn and Facebook, but probably not Snapchat and Instagram. On the other hand, if you are in the consumer IoT space, you will want an active presence on all possible platforms and forums.
3. **Engagement**—All social media platforms are fundamentally designed to be places of interaction. If you use them only to share information or advertising, you may end up having limited followership and engagement. So when putting up posts, try to invite for comments; think of introducing some gamification concepts in your social media marketing activities; and so on. Some companies go overboard with engagement to the extent of pestering, avoid those as well. Always remember this is a 'social' platform, just digital instead of a physical space.
4. **Content**—It is critical to pay a great deal of attention to the content—what you put out and how you do it. Once you share something on social media, it is out there forever. People will refer back to your content from years back to prove a point that may not even be relevant now. While it is impossible to think of all such possibilities, be very measured in the messaging which must be in line with your strategy. There should be only a select handful of people authorized to put content on your behalf so that you can control quality. Keep it as relevant to the business and your audience as possible, e.g. posting pictures of private parties on LinkedIn is highly inappropriate. You should also pay attention to the language and tone of the content so that it is in line with the image you want for your business. Having sufficient visual content and infographics helps draw people's attentions.
5. **Support**—Managing social media effectively is a lot of work. On one hand, you get a lot of traffic which is difficult to monitor and manage; on the other

hand, there is an implicit expectation that you will respond to each feedback stream. People want to their feedback to matter, unless you acknowledge and keep providing updates, they get disconnected from you. For managing responses, while there are some AI tools available, you cannot rely entirely on them; human touch is critical in developing a relationship with customers and constituents.

6. **Personalization**—This is hard because this is a mass media. In addition to making sure everything gets responded by having the right support mechanism, you need to personalize as well. You start by making the dialogue around your social media campaigns as interactive as possible. Once individuals get drawn into the dialogue try and identify with them personally. Do a little research about the people before you engage, so that you avoid any faux pox and make them feel valued.
7. **Sentiment analysis**—You should have mechanisms in place for early detection of sentiments expressed on social media, they have a way of catching up like wildfire. There are semantic based tools and algorithms that use keywords for the basis of analysis. Such tools might be limited in interpreting contextual and emotion based sentiments which have a more powerful impact. Nevertheless, whether you use sophisticated tools or not, you need to have a program to do sentiment analysis on your social media channels to understand how your customers feel about your products, services, and company or those of your competitors.
8. **Synergy**—You will most likely have multiple social media channels. You need to have a plan to drive synergies across those platforms and also across other marketing efforts in the organization. You definitively will have an online presence outside of your social media network, which also needs to be in sync. If your platforms or messaging seem out of sync, you can draw ridicule. People are less unhindered in social media as there is less human contact.
9. Branding: Similar to synergies, you need to have a consistent branding approach across various social media platforms that is in line with your overall branding mission [4]. Due to inherent differences in the various platforms and their constituents, it is easy to have deviated from the plan; but inconsistencies will impact your perceived value amongst customers.
10. **Hashtags**—They are very useful to drive traffic, themes, and content. But indiscriminate usage of multiple hashtags around the same thing can destroy the impact.
11. **Metrics**—Rigorously track the metrics around your social media platforms—number of unique visitors, the number of visits, demographic analysis of the visitors to the extent possible, interaction metrics, lead generation, etc. You need to watch out for trends and have remedial actions if required. For example, if your visitors are doing down, you may need new content or new ways generate interest in your pages. If your lead generation is going down, maybe you need to sharpen the value articulation of your offerings. You have to track what is working and do more of those; you also need to track what is not working and remediate the impediments.

Given the significance of social media marketing, you must always have a contingency plan for any major issue. Issues could be a major product defect, service failure, cyber security attack, inappropriate action or statements by any employee, or anything which can negatively impact the sentiment around your business. These things do happen and we need to plan for them. As we build disaster recovery plans, you need to have a social media contingency plan all the time. This plan should include the following:

1. Possible events and scenarios
2. Constitution of response team and their responsibilities
3. Repository of messages and responses for various events and scenarios
4. Partner and ecosystem communication protocol
5. Impact analysis methods
6. Escalation path
7. Communication channels, frequency, and plan
8. Incident monitoring mechanism
9. Executive engagement protocol

You need to conduct periodic drills to ensure effectiveness and adequacy, just like you will do for a building fire drill. A social media fire can easily destroy your business and do it very quickly.

13.6 Summary

In this chapter, we have discussed what makes marketing for IoT businesses and products unique. You need to have new approaches and use different tools for successfully marketing IoT businesses, we have discussed some of those here. We also introduced you to the 3P framework to complement your organization's regular marketing effort by engaging the people involved with the business more. If your business has multiple facets and IoT is just one of them, you need to ensure there is harmony between the marketing efforts for your IoT side of the business and remainder ones. In dealing with IoT, you need to think of marketing at the same time as you are thinking about the product. Consequently, you need to integrate the product, business, and marketing strategies into one converged plan. When working with partners, it is always good to collaborate even on the marketing front and create a multiplier effect. Social media has one of the most significant impacts in our daily lives and businesses, winning in this theater is critical to the success of any IoT business, later in this chapter we discussed some best practices around this.

We end this chapter with a quote from Joe Chernov "Good marketing makes the company look smart. Great marketing makes the customer feel smart."

In the next chapter, we will deep dive into another related and very interesting topic monetization. As you will see, both are very tightly integrated.

References

1. (Gong 2016) Gong, W. (2016). The Internet of Things (IoT): what is the potential of the internet of things (IoT) as a marketing tool? (Bachelor's thesis, University of Twente).
2. (Peter 1999) Peter, J. P., Olson, J. C., & Grunert, K. G. (1999). Consumer behavior and marketing strategy (pp. 122-123). London: McGraw-Hill.
3. (Elizalde) "A Product Management Framework for the Internet of Things" by Daniel Elizalde https://techproductmanagement.com/iot-decision-framework/
4. (Lindstorm 2005) Lindstrom, M. (2005). Broad sensory branding. Journal of Product & Brand Management, 14(2), 84-87.

Chapter 14
Monetizing Your IoT Efforts

14.1 Introduction

Every technology hype-cycle [1] brings along with it the promise of substantial economic benefits and opportunities. Similarly, IoT is expected to create new opportunities worth billions of dollars and lead to benefits worth trillions. It does not matter which numbers we believe, one thing is for sure—the impact of IoT is big. When such projections are made, the premise is always based on assumptions of how IoT (or for that matter, any other technology) is changing how we live, work and engage with our broader environment. Projections are realized by appropriate monetization effort for each individual initiative. So it matters how you monetize your IoT efforts, as the industry does not change or move without your contribution. Moreover, any IoT initiative, like any other technology initiated or technology enabled action, requires investments. The relatively depressed economic climate over last several years has strained investment bandwidths. The window of expectations from any investment is shrinking, and appetite for large-scale investments is reducing. These factors demand more focus on monetization efforts for your IoT initiatives.

To effectively monetize your IoT efforts, they must lead to greater benefits for your customers and your company. IoT enables new non-traditional methods of transacting business and creating value. We see an increase in utility-based and outcome-based business models being driven by the IoT revolution. So how you monetize matters, and this, in turn, impacts product and service design around IoT. At the same time, you must guard against competitive disruptions. This chapter helps you understand monetization opportunities and programs while avoiding common pitfalls.

An underlying dimension of IoT monetization is data. We have tackled the topic of monetization of data extensively in Chap. 6 of our earlier book 'Making Big Data Work for Your Business' [2]. The frameworks discussed there are still valid in the

S.R. Sinha, Y. Park, *Building an Effective IoT Ecosystem for Your Business*,
DOI 10.1007/978-3-319-57391-5_14

context of data. Here, we will introduce you to some concepts, models and approaches beyond data.

We cover the following topics in this chapter:

1. Understanding how IoT creates new value and business models
2. Adapting to unanticipated opportunities
3. Developing a framework for determining value of data insights and outcome
4. Organizing your business differently to leverage the new business opportunities
5. Planning to safeguard your current economic models while pursuing new business models

14.2 Understanding How IoT Creates New Value and Business Models

Broadly speaking, your business is selling products and/or services. We see most manufacturers interested in providing services around products they make for lifetime profit and customer relationship potential. Over time, value added services start getting defined around existing product/service through improved outcomes, better organization of capabilities, and/or implementation of technology that transform the nature of the product or service. The other phenomenon which transforms businesses is driven by the convergence of categories of products/services. These changes are sometimes leveraged in the primary relationship between the manufacturer/service provider and the consumer and sometimes leveraged by the intermediaries. For the rest of this chapter, we shall consider service as a product to simplify things.

To summarize, transformation leading to new value creation happens in three different ways. Let us explain with some examples:

1. **Enhancing product value**—White goods like washing machines, dishwashers, refrigerators, etc., which have been in use for decades, have become more value added with the introduction of new features leading to ease of operation, energy savings, and versatility. While the product capability might have changed over time, the category or delivery channels have not fundamentally changed.
2. **Improving organization of delivery channels**—Our next example is one where the product outcome has fundamentally remained same, but the delivery channels have significantly changed over time, leading to industry formation and shifts. Let us take the example of fitness programs. Decades back, fitness was not a large industry. You had the option of seeking boutique expertise around your chosen method of fitness programs like Yoga or Pilates, or some other form, but you would rarely see organized mass scale efforts around those. Gymnasiums and studios started forming over time to address emerging customer needs around fitness—initially as one-offs, eventually leading to large national chains. These delivery channels did not address the needs of people who have time

constraints or feel shy in public engaging in fitness programs. So some smart practitioners came up with the ingenious idea of offering VHS cassettes, followed by CD/DVD, and now web streaming personal fitness programs in the privacy of your home, in your own time, for a very affordable cost.

Next, let us look at watches, which have crossed category boundaries—from mobile mechanical time piece devices to luxury items to multi-purpose mobile electronic devices.

3. **Converging categories**—Recently, Fitbits did something interesting. It brought the product capabilities of personal activity tracking and computing to a 'watch' type form factor and capabilities. It is constantly upgrading capabilities and usefulness to upsell, fortify relationships, and to compete with mobile phone manufacturers who are embedding fitness tracking capabilities in phones.

So you can see the product and channel transformation, and category convergence changing the same swiftly and severely. This is how IoT begins to redefine industries and value creation. Actually, the impact is quite deep on human behavior. On a lighter note, I (Sudhi) have been trying to lose weight for a long time (food and laziness are my nemeses). Every now and then, I will exercise. Now sweating is not good enough—my Fitbit or phone has to sing praises of my awesome efforts. However, I continue to buy larger size clothes and belts.

The potential of technology and blending of services/products are not fully exploited yet. You can marry the Fitbit type devices with actual training programs (Yoga/CrossFit/regular gym/Zumba), custom food services (Jenny Craig), and advanced analytics to create a more holistic and personalized fitness program.

Traditionally, we have required assets to possess an explicitly defined and generally accepted value to be worth something for any financial modeling. This paradigm transforms significantly in the world of IoT where every 'thing' is an asset, whether it has a value today or not. Even if the 'thing' does not have a value today, the data it generates or the outcome it causes might acquire a different context, and thereby, value at a future date. By enabling interaction and engagement, in a way IoT brings life to 'things'. So our window of value consideration changes from the narrow immediate one to a broader lifecycle one. Value is generated at every step of the way in the IoT ecosystem framework discussed in Chap. 1. Value is also generated over time through the life of the asset and associated transactions. This is the basic premise of how IoT starts to redefine value.

From the beginning of history, when we started putting a value to things or transactions, the extent of value was grounded in demand/supply equation or desire equation or some combination of these. Need, want and availability still drives the perception of value. These manifested in extrinsic value, which over time started creating intrinsic value. This is how gold became more valuable than silver or iron. In some cases, the outcome had a tertiary role to play. When it comes to services, similar considerations drive value. In almost every case, the extent of value diminishes over time due to commoditization and lack of substantial change in the impact of the product or service—exceptions are cases where availability continues to shrink, or exclusivity continues to increase.

The world of IoT operates differently where perceived outcome and its impact takes a primary role in determining valuation, while other factors become more of a supporting cast. For example, accelerometers have been around for a while. The concept is nearly 100 years old. Over this long time horizon, we had many implementations of the technology including activity tracking and step counting. High sensor cost and low market demand were factors limiting mass consumer adoption. Fitbit, which can be considered an IoT instantiation of the same technology, amplified the value. Now, people perceive it to be a critical element of healthy living by accurately tracking all types of physical activities and behaviors.

So you need to think of value in the IoT world both in context of lifecycle and the outcome, transposed with user engagement or transactions over time. It is difficult to separate the physical product and associated outcome or service. Jet engine manufacturers trying to sell thrust, large capital goods manufacturers trying to sell products, and lifetime predictive diagnostics device manufacturers trying to sell emergency monitoring and response services for elderly, are all examples where IoT-enabled products and saleable services are intertwined. In all these cases, the value will have to be qualified in terms of customer's willingness to pay (defines the quantum of value) and permission to sell (defines readiness to capture value).

When thinking of IoT monetization and value creation, you need to consider a few fundamentals:

1. **Growth curves**—Over time, your volume of connected products, customers, and services delivered will be expected to grow. This creates your growth curves. Like any other product or business, your interventions like marketing efforts or product enhancement, market adoption, customer perception of value, customer feedback, etc., will change the direction of the curves. IoT enables you to change your engagement models and paradigms with your customers, so you need to plan the interventions very well. Keep in mind—technology is changing, user preferences are changing faster, entry barriers for the frontal or fractal competition is low, the real and perceived value of your product is also changing [3]. So you need to thoughtfully and deliberately plan for your growth curves and interventions to influence the slope of growth curves. Sometimes, planning for declining slope over a period of time might also be part of strategy if that subsequently helps improve volume (Fig. 14.1).

   Typically, businesses consider volume growth in terms of a number of customers, units sold, and/or revenue. Dealing with IoT, you need to additionally factor in the growth of connectivity, data, and derived services/products. Keep in mind that often time-to-connect is a function of market and technology maturity. While the cost of storing and processing data has significantly come down, the cost of connectivity in certain use cases might still be a limiting factor (at least in certain markets). With the advent of 5G and other improvements to communication technology like wireless, Wi-Fi, etc., we will see greater opportunities to connect more people and devices.

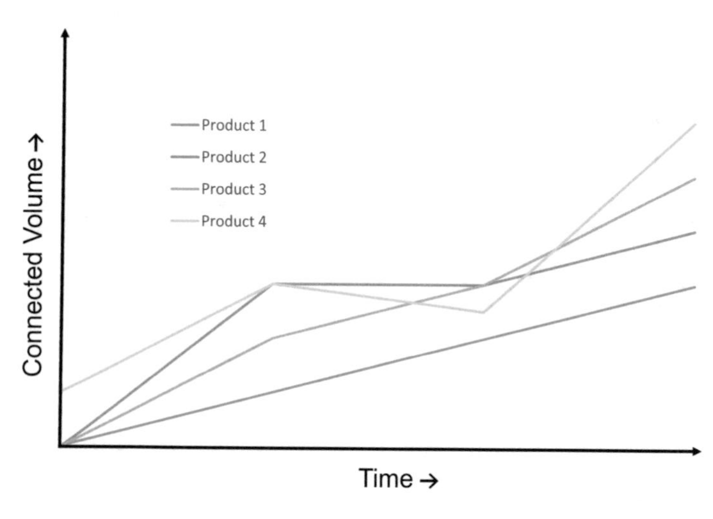

Fig. 14.1 Growth curves

2. **Discount factors**—Factors which drive value for your products/services may have a diminishing effect over time. The most common factors experienced by traditional businesses are declining input costs, competitive pressures, and delivery channel efficiency. Managers are always trying to optimize cost and value equations so that profitability is not negatively impacted. In many cases, companies are successful in managing the factors, or there is adequate opacity of the market that the impact of these factors on perceived value and price is minimal. In the IoT world, the rapidly declining cost of storage and electronics has already established expectations of customers around reduced price of product/service. Additionally, you will frequently face disintermediation from non-traditional competitors (channel transformation or channel convergence) that will impact perceived value of your product/services. The outcome that the customer is paying for might also have diminishing returns over time. For example, if you have an energy savings initiative using sensors and analytics—beyond a point, your ability to save the same amount of incremental energy will go down; if you are trying to achieve health goals—once you have achieved your goals, your willingness to pay increasing amounts will be less (Fig 14.2).

   In designing your monetization strategies, you will have to factor in all these reasons the value of your product/services might diminish over time. Modeling these discount factors and their impact will give you more insights into risk, which in turn will help you be more predictive about your business and its long-term value.
3. **Economic profit and growth based valuation methods**—When building your IoT-based business, consider some of the traditional methods of valuation as well. Professors David Wessels, Marc Goedhart and Tim Koller in their seminal book "Valuation: Measuring and Managing the Value of Companies", have

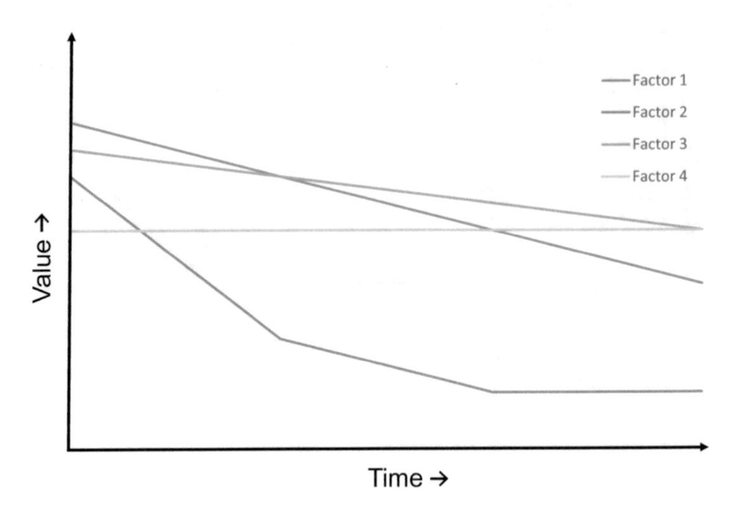

Fig. 14.2 Discount factors

extensively discussed the different methods and formulae for such calculations—a very good read for any current or aspiring business leader. In this book, Prof Wessels and his colleagues explore the impact of various financial parameters like invested capital, return on invested capital (ROIC), weighted average cost of capital (WACC), discounted cash flow (DCF) among others. One of the key factors that impact valuation is how growth alters the valuation equation in the context of these various financial parameters.

Some of these popular calculations will come handy for you:

$$\text{Economic Profit} = \text{Invested Capital} \times (\text{ROIC} - \text{WACC})$$
$$\text{Value} = \text{Invested Capital} + \text{Economic Profit} / (\text{WACC} - \text{growth})$$

Growth is defined at the rate at which net operating profit (after taxes) and cash flow grows annually. ROIC equals net operating profit after taxes divided by invested capital. WACC is a standard rate at which a business can borrow money.

To ensure that you have better economic value for your investments and initiatives, your efforts must be directed at the following actions:

1. Keep investments as low as possible. If investments seem to be very large to make your initiative a success, try to partner with your processing chain participants and customers to share the investments.
2. Secure revenue as quickly as possible.
3. Focus on returning a profit income as early as possible.
4. Steadily grow the business—stagnation destroys value.
5. Get as much predictability as possible around your operating and financial parameters.

14.3 Adapting to Unanticipated Opportunities

IoT is growing on the back of rapid technological advancements, and in turn, IoT is transforming every possible business. It will be suboptimal for any business to have static monetization models in this rapidly evolving flux, so you must be able to adapt. Let us explore a few of the countless examples of unanticipated opportunities.

The high volume of transactions and diversity of use cases often lead to platform-based businesses, which is fast emerging as a new segment. Several companies which were into traditional products and services are now using their platforms to build completely new business around these platforms. General Electric (GE) has a long history in industrial automation starting with supervisory control and data acquisition (SCADA) systems for large equipment, and systems like jet engines, locomotives, and other industrial equipment. As technology evolved, Predix was born as a platform for data analytics and control for large industrial equipment. The initial goal was the effective and optimized operation of large industrial equipment, predictive diagnostic and maintenance, and the ability to offer new value-added services to customers. Over time, GE also explored new business models on these new capabilities of selling performance instead of just selling a piece of hardware—for example, selling thrust instead of selling jet engines. Seeing the emerging needs of similar industries and other companies needing a robust data analytics platform, GE has now expanded into an IoT platform business using Predix.

Another large industrial manufacturing company ventured into the IoT space with an initial objective of performance monitoring and predictive diagnostics leading to improved service linkage, boost the productivity of service technicians by giving them insights into equipment health before they reached the maintenance site, increase additional time and material services while performing regular maintenance activities by pre-identifying issues observed from data analytics, and grow the retrofit opportunities pipeline by proactively reaching out to customers with rationale around early replacement (where there was substantial performance degradation of the equipment). Once a critical volume of connected products was reached, the company also saw a reduction in warranty costs, since they were aware of equipment issues before customers reported and made claims. These usage and performance insights were also used by the company to influence their future product development roadmaps.

Amazon started out as and is a very large global retailer. To support their humongous retail transactions over the web, they had to build large data centers and efficient ways of storing and processing data. These capabilities lead to the development of Amazon Web Services business, which is one of the leading public cloud services provider in the world today. Their peer in the public cloud space is Microsoft Azure. Microsoft was building new capabilities around Windows operating systems as part of the RedDog version for cloud hosting of applications like Xbox and large scale SQL instances, which lead to the development of Azure.

IBM has been very keen to develop a substantial data analytics based business. They have invested in very sophisticated capabilities around machine learning, deep learning, and artificial intelligence among others in their Watson platform. Now they have acquired the Weather Company to leverage the changing weather information for their clients, and also create new offerings by analyzing the impact of weather on different businesses and communities.

These type of unanticipated consequences happens with much smaller companies as well.

Augury is a predictive maintenance company that has developed technology for acoustic signature analysis of industrial machines. The analytical capabilities and algorithms developed by Augury are now being leveraged to extend their platform beyond acoustic study to broader areas of maintenance. Their focus has shifted from their previous core strength of hardware to data analytics.

Similarly, another Silicon Valley venture Predii started with a focus on repair diagnostics for automotive components. Their first major customer was SnapOn who used their platform and analytics to improve their productivity and revenue base. Now, the Predii platform is being extended to other devices and systems with a goal of becoming a leading player for Repair Intelligence™.

Such examples are countless. For smaller start-up type companies, adapting to evolving business models and revenue streams is often easy—for more established companies, refining their business models is sometimes difficult. But the key factor is, whenever you are building an IoT-based business and are working on monetization strategies, you need to constantly review and ask yourself—Are there other avenues to create value for customers and capture that in the monetization framework?

Exploring these unanticipated consequences periodically, systematically and frequently will help you reverse the impact of declining growth curves and discount factors. To programmatically tackle this, ask these questions as part of your operating cadence:

1. Are we helping customers solve problems beyond what has been promised?
2. Are there new ways we can combine existing data to get new insights? Can we turn those insights into something useful for customers?
3. Do we see opportunities for new classes of customers?
4. Are we adding new capabilities to our product, service or technology platform that might be of use to other companies?
5. Is there an opportunity to collaborate with other companies or entities to provide more value to customers?

Unanticipated consequences create a lot of excitement and possibly good profit opportunities. They also create tension in teams and organizations. Often businesses or initiatives, especially smaller ones, have the tendency to divert from their initial charter once they experience these new opportunities. It is critical to assess at such points the impact of monetization efforts from your original efforts, draw out the growth curves and discount factors for original as well as new opportunities, evaluate the economic profit and valuation—first independently followed by jointly, and then make choices on how you want to pivot your business in future. The timing of making any pivotal changes will be crucial—you have to evaluate the technology and market adoption rates, as well as evaluate customers' willingness to pay and your permission to sell.

14.4 Developing a Framework for Determining Value of Data Insights and Outcome

While technology changes fast, businesses and economic models take longer to transform. Such transformation is usually associated with infrastructure and behaviors, which take the time to change. Consequently, we often see the influence of some traditional thinking in developing economic models around IoT. At the outset, we would like to address a few of these common mistakes and myths:

1. **Return on Investment (ROI) should be the starting point for any value capture**—The ROI approach obliges us to think of investment or cost side of the equation early on. While it is very important to be prudent about cost, the first focus should be on how you are improving Return on Asset (ROA) for your customer. This will compel you to focus on your solution that will deliver outcome improvements for your customers. Sometimes you may not have the right metrics to make an undeniable value argument with your customer, but if you collect enough data and do it for a period of time, you will be able to identify the value equation for your customer. For a number of IoT use cases, especially in the consumer segment, thinking in terms of physical assets or ROA may not be the right answer; consider productivity improvements or other returns for the value equation.
2. **If data is an asset, we must own the data to capture value from it**—Since data is the new currency in this increasingly digital and connected world, there is a race for ownership of this new gold. While data has intrinsic value, it derives its true value from inferred implications. Ownership of data can at times become a very contentious topic with your customers and other participants in the IoT ecosystem. Customers or owners of the assets or transactions that generate data will claim ownership of the data; participants who are transforming the data in the processing chain will also try to establish a similar type of ownership, as they claim to be 'manufacturers' of this newly interpreted data. Over time, data will change context and acquire new value, and then you will have a new set of claims to resolve. When asset and transaction generator rights change hand, does the data ownership also naturally change hands? There are too many factors to consider. However, this debate should not slow you down. What you need is free and clear access to use the data and process it further in perpetuity. You also need protection that in case your customer enters into any agreement with any third party, your rights to use the data will not be impacted.
3. **Voice of customer provides the guiding principles for value capture**—Most of our prevalent marketing and business design disciplines are centered on feedback from the customer. While this is a very useful and practical tool, history shows us that for transformative results, it is not the most effective. A very famous quote attributed to Henry Ford, "If I had asked people what they wanted, they would have said faster horses", from nearly 100 years back, reflects the latter philosophy. In more recent times, Steve Jobs and many other great innovators have been seen to follow a similar approach. If you are bringing an entirely new class of product or service, it is very difficult for customers to use a good existing benchmark to give you better comparative insights. Also, if your product/service

is expected to transform rather quickly and you cannot gain the meaningful voice of customers too frequently, it becomes cumbersome and expensive. So it is important for you to find the right balance between how much you want to depend on the voice of the customer for value capture. Your conviction about a strategy and a path must be rooted in practicality and some form of validation is useful—but, balance it against uncritically the following feedback when you are trying to fundamentally change some paradigms.

4. **The business will not become profitable before reaching full scale**—Most IoT businesses if designed right, should be quick and require low investment levels (this is always relative) to get started, get some new customers, and try a few use cases to see if you can create and deliver value. Yes, you need to scale to realize the full potential of your business. But, if you cannot make your business work on a small scale (gain a critical mass to sustain and grow), you are most likely doing it wrong. It is quite possible that you will be disintermediated or displaced by a new player quickly, you may have to change your target customer base or business model to better adapt to the emerging value capture or competitive scenarios, and the technology or user preferences might just change over time—IoT businesses are subject to more dynamism than traditional industrial or consumer businesses. So you need the flexibility to scale up or down quickly without compromising your basic business and technology platform—early break-even is critical for a successful business design.
5. **We can get really good at generating data insights based on our generic analytical capabilities**—Domain contextualization is very important—data in its native digital state does not have any value. It derives value from contextual semantics. It is proven that you need advanced statistical and analytical skills to discover substantial value from data, but analytical skills alone cannot accomplish the goal if domain contextualization is absent. When you rely primarily on generic analytical skills, you will get good at finding the 'what', but you will have limited opportunities to interpret the 'why'. Without understanding 'why', you will not be able to ideate on 'what to do next' and 'how'. In our experience, it is incredibly hard and time consuming for multiple generic analytical platforms and services to build a comprehensive semantic domain library, as it takes years of know-how to build it. Due to lack of sophisticated knowledge management systems in most organizations, a good portion of these insights reside as tribal knowledge in people's heads. So, focus on building a strong understanding of the domain as you start building your expertise on data.

As discussed earlier in this chapter, we will not rehash the whole topic of monetizing data—for that, please refer to Chap. 6 of the book '*Making Big Data Work For Your Business*'.

In the world of IoT, product and service are entwined. This is important, as you attempt to build a long lifecycle relationship with your customer and capture value throughout the lifecycle. In this section, when we talk of 'product', we are really talking about the sale of a physical product that is establishing your customer relationship or the first instantiation of a new service class offered to the customer. Our definition of 'service' in this section is the derivative transactions from the sale of the 'product'.

There is value capture from transactional interactions around the connected product itself. You have the first cost of buying or installing the product, and you may have some periodic subscription type charges for the service associated with the product. In determining this transactional value capture, you need to consider the following:

1. Real benefit
2. Next best alternative
3. Perceived value
4. Desirability premium
5. Convergence premium

There are additional value capture opportunities for the performance-based outcome from the service offerings around the product. You also have service cost optimization opportunities for the end customer and the intermediaries you often encounter in IoT ecosystems. For example, if you can lower the maintenance cost for an industrial equipment by using data analytics from the IoT-enabled connected equipment, you can leverage that for monetization opportunities either through a maintenance service provider or directly to the end customer/asset owner.

There is money to be made in the aftermarket ecosystem as well. Once you have the customer hooked to your connected product and services, you can use data analytics to identify opportunities for parts, accessories and additional services. You could also use insights from your customer and installed base to enable other partners to offer similar parts, accessories, and services; this is similar to providing brokerage services—bringing other providers to your customers.

There are new value capture opportunities that will be created by the data insights. You will be able to enhance the value of your offerings or create new ones, and deep analysis of customers' usage of your IoT products and services will help you identify latent needs or lead you to unintended consequences. You can also sell data insights to ecosystem business partners to be able to create similar new opportunity landscape—this type of brokerage service brings your customers to other providers.

You also have monetization opportunities from new offerings that can be created from scale economies. Examples are demand forecasting and supply chain optimization.

Table 14.1 captures the various value sources we just discussed.

In most traditional businesses—industrial, consumer or service-oriented sectors, monetization and value definition or value capture modeling is carried out in long intervals. This is because the products and services are not refreshed very frequently. In the IoT economy, it is different. You get new insights and find new opportunities, so you need more frequent refreshes of your offerings and business models. Build in a cadence to review the value capture framework quarterly as part of your operating rhythm.

Table 14.1 Value source analysis

Value capture source	Value driver
Product and service	Traditional
	Performance outcome
	Service optimization
Aftermarket ecosystem	Parts, accessories, services
	Business partners' parts, accessories, services
Data insights	Feature enhancements
	Customer insights to business partners
Scale economy	New offerings, new business

14.5 Organizing Your Business Differently to Leverage New Business Opportunities

You are either a new start-up or a well-established business trying to leverage IoT. If you are the former, you are likely to review existing organizations and business models in same or similar spaces to draw inspiration on how to organize your business. You will take the basic fundamentals of existing companies and adjust models to compensate for differences caused by IoT. If you are an existing well-established business, you are likely to take a similar approach. You will look deep into the history, performance and experiences of your business to understand the strengths and improvement opportunities for business design—compare it to those of your competitors, learn from players (like start-ups) who have forayed into your space, and then finally arrive at what you think is the best model after accounting for the IoT influence.

As we have seen so far in this chapter, and as we see around the world every day, IoT brings new non-linear business opportunities—whether it be new products (Fitbit), new service offerings (outcome based services), new segments you serve, or entirely new business models (thrust as a service instead of selling engines—Uber). These are fundamentally transformative. Applying traditional methods of customer relationship management, pricing, channel development, marketing, product development, sales incentives, and other disciplines involved with growing a business, can become suboptimal for growing an IoT-based business. Let us consider some scenarios to understand this phenomenon.

Example 1 Start-up with an integrated product and service for B2B market
The company makes sensors to understand the mechanical vibration and electrical signatures of large industrial equipment. The sensors have some storage and communication capabilities to be able to temporarily store or cache the data and send it to some sort of central cloud environment for long term storage and processing. This company has developed proprietary advanced algorithms to interpret the vibration and electrical signatures leading to understanding fault potential and possible limited diagnostics of the industrial equipment the sensors are attached to. The company makes these analyses available to users, maintenance crew, and manufacturers of the equipment for a price. This is a typical IoT enabling business.

In the traditional model, the company should maximize the profit on the sensor hardware, explore opportunities to create revenue stream around data management, and sell the analytics for a reasonable fee. The company should develop multiple channels to sell their equipment and services with a high focus on the asset owners and maintenance companies because they are the most obvious beneficiaries. Product development efforts should continuously strive to drive sensor costs down, pack more power in them making the analytical platform more user and customer friendly. Sales incentives designed around percentages of the value of sensors and services sold sounds like a valid strategy.

However, the company realized that this approach is somewhat shortsighted and flawed. Sensor costs will keep going down. Achieving extraordinary performance from sensors will require massive sustained R&D investments which will be challenging for a start-up environment. Since there are millions of asset owners and consequently maintenance touch points, trying to sell to this vast market will be very expensive and time-consuming. Vibration and electrical signatures are two inputs to understanding equipment health, and there are other factors. So the company chose its growth strategy to be centered on the analytical platform and capabilities, as that is where maximum value capture opportunities reside. So this company needs to get access to more different types of equipment performance data—preferably data generated by the equipment itself. While there are millions of asset owners and maintenance providers, there are a limited number of manufacturers of expensive complicated industrial equipment. So the company decided to pursue a partnership with the OEMs and complement the value provided by the asset and OEM services around it. The company shortlisted OEMs who have a similar goal of exploiting sensor data for improved maintenance and started working with those who have an active initiative in this space. Instead of requiring OEMs and customers to always use their data analytics platform, the company has flexible and often reciprocal relationships of using others' platforms. Instead of investing in a significant sales force, the company invested in partner development organization incentivizing access to large install bases of OEM equipment.

The end goal of the company is to understand every piece of data that influences maintenance for large equipment, develop the most comprehensive analytical algorithms, and have relationships to be able to access a wide array of data. Traditional approaches might have led to some missteps.

Example 2 Large industrial equipment manufacturer serving the B2B market This company makes capital intensive equipment that lasts long and is often at the core of the asset owner's business. These equipment are very complex machines with mechanical, electrical, electronics and software components embedded into the products. This company has few competitors but high growth and margin pressures. It makes less margin in the initial sale of their products but has a very attractive parts and services aftermarket business around the installed equipment. The service business comprises of standard contracts which have blocks of technician time to perform a list of maintenance tasks similar to the 3-month service plan for cars, and other value added services available at a price. The company has impressive sales channels, strong product development, and manufacturing footprint, distributed and very experienced service teams and regional parts hubs to manage aftermarket logistics. This is a typical traditional business.

Because of the weak economic environment, the company is facing sluggish demands for new products. Since it makes very good and reliable products (which helps with initial sale and subsequent reputation), there are limited opportunities for replacement equipment. The biggest growth opportunity for the company is to aggressively go after whatever is left untapped for the aftermarket service business. To achieve this goal, the company needs to establish service contracts with customers who do not have one. To do so, the company needs to demonstrate an increased value from such service contracts—giving customers periodic or real-time insights into equipment health and performance through remote access is one way to increase service contract appeal. The company also needs its customers to make more use of its value added services like executing proactive maintenance protocols. The company needs to understand the full potential for replacement parts and prepare their supply chain organization to meet demands.

To accomplish all of the above, the company has initiated a program of connecting their equipment, collecting all sorts of sensor data from the equipment and other environmental data which impact equipment functioning, running advanced analytics on a cloud-based platform leading to health monitoring/reporting and to the identification of predictive diagnostics. The company is leveraging IoT to accelerate its service business. However, the company is slowly learning that its service business is transforming. Since the company can do predictive diagnostics, customers are less inclined to pay for a block of fixed hours for a set of routine maintenance activities. Customers have started expecting access to their equipment health and performance monitoring without paying a huge price. The service technicians who relied mostly on their extensive experience may now be receiving different inputs from the analytics which is non-intuitive to them. You can get better technician productivity, as you can arm them with a lot of equipment information before they reach the site, but your union contracts might interfere with your productivity capture efforts. The supply chain and logistics organization are designed to keep trade working capital (TWC) low, thereby driving least possible inventory. However, to fully capture replacement part opportunities quickly, you will need to balance your TWC goals. Until you get a large enough population of connected equipment, you will not be in a position to run machine learning or any meaningful level of aggregated analytics. So you have to keep connecting until a point, possibly without seeing any return. Customers do not want to pay immediately because the value is not obvious or established for them—and your traditional business case methods find return-less scale-up difficult to accommodate. In the traditional model, it was critical to have better and more expertise in the field, but now you need to balance that with expertise and scale at the central service bureau which is driving this connected program.

Traditional companies with large installed bases and service workforces can find it difficult to adjust to this type of service transformation. However, if you do not model the scenarios and their implications well, you will not be able to take full advantage of your investments.

Example 3 Manufacturer of a wearable device serving the consumer market The company makes popular wearable devices like watches. The company is recognized for its strong brand, quality, product accessibility (channel presence) and value pricing. The company undertakes massive marketing efforts to keep its brand recall relevant. The company frequently refreshes its products and comes up with new ones to keep up with changing consumer tastes. This is a very typical consumer company.

The company is facing a few challenges. New competitors buoyed by reduced manufacturing costs and larger investment portfolios have begun to chip away market share. It is becoming very hard to predict the brisk pace of consumer preference changes. Other manufacturers from different categories have started to incorporate this company's product functionality into their offering, further destroying market attractiveness—like people often use phones as timepieces instead of watches. The company needs a new business and product strategy with revised business models to stay relevant.

Due to rapidly reducing sensor costs, data transport costs, and data storage processing costs, the company has decided to pack more into their products and extend their current business model. By monitoring activity and personal health parameters (several of which can be determined by the new sensors put on the product), and making this information available to the consumer through phone and other apps, the company expects to attract new customers, improve the utility of their products (and consequential value), offer additional services, and open up a whole new slew of possibilities. The company has toyed with the idea of using all the new information it is collecting to compute people's health risk index to influence insurance premiums. This will allow the company to become a broker of services based on the information it is collecting—it can help customers pay reduced premium on account of a healthier lifestyle, and it can provide insurance companies a channel to offer new services. Now, this company is also trying to transform by riding on the IoT bandwagon.

There are a number of unknowns with this new direction. In the traditional model, the sales channels were incentivized by the commission, which drove pricing and promotional strategies. Now, the goal is to drive revenue and value through data based services derived over a longer horizon, so the incentive models do not apply similarly. Customers are used to making one-time investments whether in watches or mobile phones—if they are required to pay subscription fees to be able to access new capabilities, their propensity to buy might be less. The entire product development focus and organization needs to change because the product itself is changing. Previously electronics, hardware, and industrial designers had sway over the product now, user experience specialists, data scientists, and research designers also have a significant influence. The product development team will have to be complemented with a robust data analytics team. Since social media has such a significant influence on our lives these days, the company has to have new strategies, consumer engagement models, and workforce to address the implications. Though using the health data to influence insurance premiums seems a very attractive and valuable service, customers may have objections due to concerns around

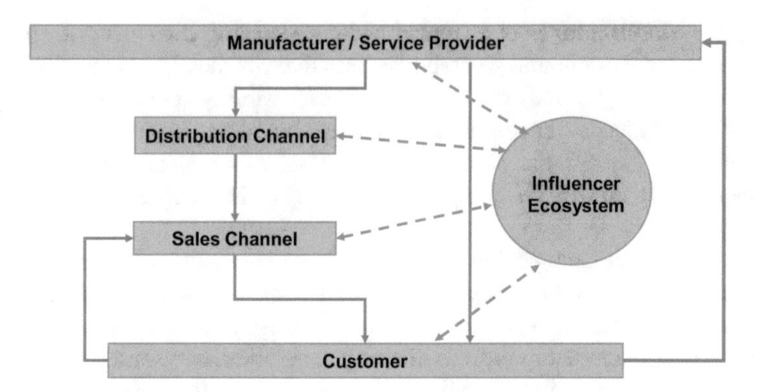

Fig. 14.3 Value chain analysis

privacy and feeling of control passing over to big insurance companies. The negative perception around such concerns has a dramatic impact on sales of the device even for its original purpose of intended use.

To leverage the power of IoT, the company has to rethink how it makes products, how it goes to the market, what it values and how it reacts to consumer perceptions.

Now you see that you need to organize your business differently than has been done in the past to take full advantage of the possibilities of IoT. To do so, you need to follow multiple steps:

1. **Value chain analysis**—You need to plot the entire route through which your products or services reach the customer. At every step of this route, you need to understand the economic incentives and business drivers. This type of process has been traditionally used for a long time to organize businesses and design operating (including economic) models. Companies form functions and departments to support the various tiers of this value chain. In the IoT ecosystem, this activity is still your first step. You can modulate the various levers that drive the economics and actions of the different tiers to influence customer relationships and business outcome. Marketing and promotional campaigns are one example of such a lever. Nearly all businesses have external factors or organizations (including social media in recent times), which influence every aspect of the business processing—product design, marketing, customer preferences, pricing, etc. Figure 14.3 below depicts a typical value chain organization.

   Now you need to overlay the IoT influences, especially value created by data analytics in this value chain. You start from the product or service, understand the capabilities of the product/service, capture the data generated by those capabilities, outline the analytics performed on the collected data, derive insights from the analytics, and finally deliver value to various stakeholders—customers, sales channels, and manufacturer or service provider. The diagram below shows the IoT overlay and interactions (Fig. 14.4).

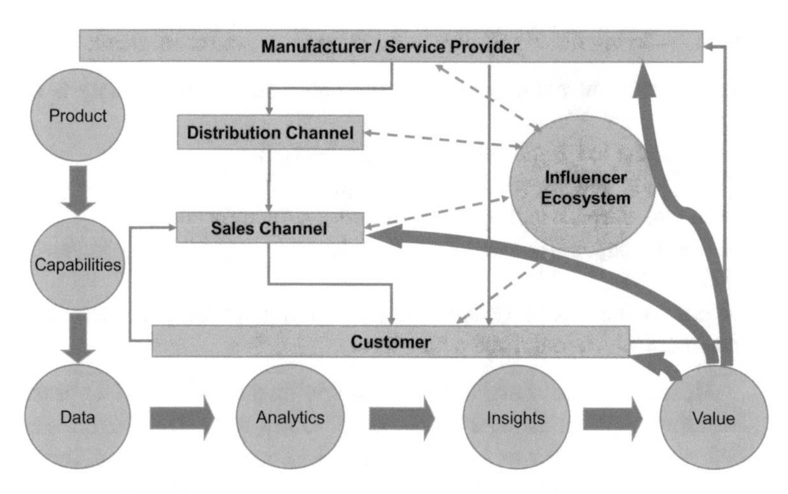

Fig. 14.4 Value chain analysis with IoT overlay

When you think of the incremental value now getting created at each step and understand its impact on the rest of the ecosystem, you will come up with more targeted and acceptable monetization strategies. This exercise will also help guide your subsequent capability design activities.

2. **Capability and organization modeling**—Now that you understand how you are creating value at every step from product to customer and back, you can easily determine what capabilities, infrastructure, and knowledge are required to harness the potential. The capabilities will sometimes be human and sometimes technological. In all the above examples, there are a few common capabilities that the companies must acquire and get very good at to be successful—for example, inexpensive data collection and processing capabilities, analytical capabilities in the platform augmented with data sciences capabilities, new service delivery capabilities, new customer engagement capabilities. For effective and quicker capability development, it is critical to acquire talent and infrastructure without entirely depending on organic build-up. IoT-based business requires different mindsets, different approaches, and higher agility, so depending on organic build-up may not get you the required results.

   Capabilities without supporting organization will not remain sustainable. For example, if you need to build a strong data sciences team, you can not apply the same employee engagement, performance management, and operating rhythm as you would apply to a normal sales or customer service organization. Similarly, if you are building an IoT-based big data platform, your DevOps function will be quite different than your normal team handling infrastructure. You need to understand the unique aspects of the capabilities required for success and then build the organization around those. This is a delicate balancing act between your current organization and the adaptations you need to capitalize on the IoT possibilities.

When modeling the capability and organization, we recommend:

(a) Assess your current organization and carry out a SWOT analysis of each function
(b) Simulate the IoT implications on each function
(c) Identify any new functions that are required
(d) Refine the purpose and operating model for each function
(e) Define the capabilities and corresponding capacity required for success by each function
(f) Determine the capability and capacity building plan through organic change management, training, and acquisition

Most companies have a good strategic talent review (STR) process, at least for the senior management levels. Usually, these STR exercises are carried out once a year and cascaded down. Success in IoT requires you to increase this by several notches—you need to expand the span and scope of STR multiple levels down, and you also need to review the design more holistically in the context of expectations, value capture opportunities, capabilities, capacity, and interaction.

3. **Incentive design**—Since IoT is changing your business design, you cannot drive the new design using old incentive mechanisms. If your goal is long-term value maximization, you need to align your reward systems to that, as opposed to a meeting near-term growth objectives. If your program is designed to incentivize big value sales with high margin dollars, your sales team will struggle to find the time and eagerness to sell low dollar value subscription services which make an impact over time. If your product development focus celebrates big launches every year, teams will be less motivated to release incremental but impactful adaptations throughout the year. The psychology behind incentives and reward systems is very interesting. We do not claim to be anywhere near experts in that discipline. However, we would like to share some insights from our experiences:

(a) People who are naturally drawn towards IoT businesses and technologies and who demonstrate a tremendous amount of passion, almost have a zeal that they need to prove to the rest of the world how they will change it. The ability to freely experiment in a contained setup and be championed in their efforts is a huge motivator. They want access to an environment where they can make a difference and they absolutely need the company of similarly capable people.
(b) People in established functions like sales, management, and even engineering in certain cases, have an expectation of additional incentives on slightly over-achieving their goals. This type of behavior is not typically exhibited by people described in 3(a).
(c) Your channel partners may not always have the incentive to drive your IoT agenda because long term some of them might get disintermediated. So you

need to design very compelling incentives for them to participate now or show them a different path for future.

When you understand the various value creation moments in your IoT implementation and understand the behaviors that will amplify the value capture, design incentive programs around those moments and behaviors. There is no unique formula yet which will work for all, or for at least a broad section of examples. Build your customized model which works best for the success of your business.

4. **Refine review rhythm**—Most businesses have very structured and regular review cadence which helps them understand the progress being made and corrective actions (if any) that need to be taken. The usual review processes look at market trends, sales pipeline, new customer acquisition and retention trends, revenue and margin goal accomplishment, progress against strategic initiatives, cost of doing business (SG&A and others), recruitment, other financial metrics like TWC, accounts receivables, accounts payable, supplier performance, service team efficiency, manufacturing performance, quality and other similar parameters which define the business. The best reviews look into future forecasts or simulations in the backdrop of current and past performance—these reviews go deep into root cause analysis and focus more on adaptive actions. The less effective reviews spend more time on report outs and some color commentary by operating managers and executives.

   Many of these metrics are very relevant for an IoT-based business. However, you need to also care for a few other dimensions:

   (a) Growth of connectivity (installed or connected customers/assets) and reasons driving that growth
   (b) Evolution and diversity of data collected, and access to new data sources
   (c) Progress in different types of analytics developed/performed and interrelationships between the analytics developed if any
   (d) Review of how the analytics are creating value for customers and the company
   (e) Evaluation of new monetization opportunities possible through the analytical insights
   (f) Feedback on new customer engagement points or models
   (g) Customer and internal behavioral or operating process changes being necessitated or influenced by your IoT initiatives

IoT is expected to transform your business, so you need to continuously and diligently review the transformative impact in addition to usual business performance. Changing your usual review rhythm is hard, and you will often face resistance from various functions because it destabilizes their sense of control. But, work collaboratively to clarify the implications and importance of the changes you are driving—the conversation has to pivot around how IoT is helping grow the business into the future.

14.6 Planning to Safeguard Your Current Economic Models While Pursuing New Business Models

IoT initiatives have a very high probability of adversely impacting your current revenue, margin and business models. Business models, customer engagement formats and frequency of interaction with products and services will change, and how value gets created and captured will also change. While you want to capture all the possibilities of IoT, you do not want to impact your current business negatively earlier than necessary.

As an illustration, in the Example 2 discussed earlier in the chapter, the service business of this company is based on linear economic models based on time and material. A Larger number of service contracts will allow access to bigger buckets of hours and a larger pool of opportunities for parts. The difference between labor hour cost and price, route optimization, effective pricing strategies for aftermarket parts drive up profitability. Successfully sustaining and growing the business is driven by a strong focus on the relationship, dependable service, and supply chain optimization.

In the connected services scenario, you have additional drivers for growth—insights, and outcome generated by the data analytics platform, perceived and a real outcome value for customers, greater productivity in sales, and execution of additional service offerings. Now, you have non-linear influences on your business. In Example 3, previously the economic model was based on branding, better pricing strategies, increased market reach, and product cost optimization. Under the IoT influence, those are still valid, but you have additional ones—new value added services, new sales models, and new customer touch points. In both examples, income from new business models will be comparatively less than the traditional ones for a period of time. If you have different parts of the organization attending to the legacy and new opportunities, there will be clear misalignment—one will influence based on today's reality while the other will influence based on future promises.

The legacy and new drivers create some tension in the organizational ecosystem. A number of large technology companies which are moving from annual or periodic license sales to subscription and consumption models are facing the same issue. Change management is hard because you have to deal with people's behaviors and motivators. So you need to plan carefully to safeguard your current economic value creation and capture models while pursuing a long-term IoT-enabled strategy. To successfully do this, you need to take three simple steps:

1. **Simulate your traditional business decline/growth and IoT-enabled growth**

   With your traditional business, there are two possibilities—either it is stable and marginally growing, or it is in a decline mode. Let us call these scenarios as Traditional 1 and Traditional 2. With the IoT-based approach, you expect to see growth, but over time. You can plot these growth or decline curves over time (as in Fig. 14.5):

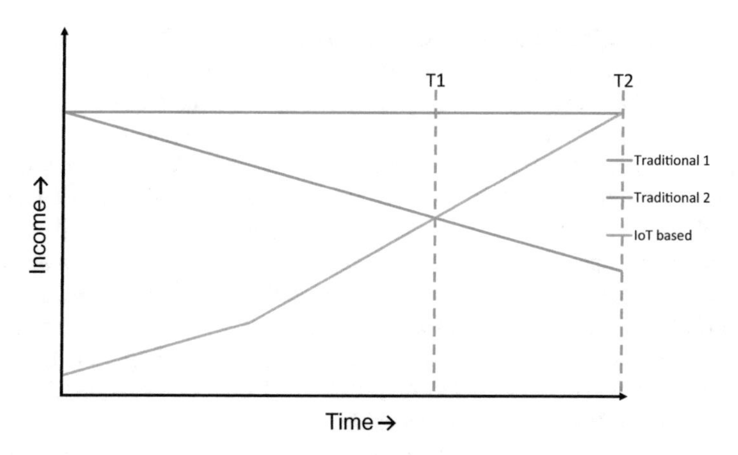

Fig. 14.5 Growth/decline curves

In either scenario, you will reach an inflection point where your IoT-based business will overtake or be the same as your traditional model. If you are not reaching out to new customers in new market segments, you will cannibalize some of your current business making the curve look more like Traditional 1. This is a very simple representation of the model, but you can make it quite sophisticated by overlaying market growth, macroeconomic factors, technology cost decline based discounts, labor rate increases (when applicable), inflation, etc. In any situation, this type of analysis will help you ground business planning into reality. Once you do this, you will understand the window for reaching the inflection point.

2. **Throttle resources to optimize the growth/decline curve**

   Now that you understand what point IoT-based models will dominate the economics of your business, you can make choices around how to amplify the growth of IoT-driven business. This can involve many things including throttling down the resources for your traditional business. For example, you could artificially increase the pricing structure for some of your products or services in a traditional model to divert more customers towards the new model. For driving more outcome based service models, you can offer some of the time-based service models as free or at a cost to create an incentive for your sales people to push the new services more. You could change the sales incentive structures to drive more long-term subscription type models. If you are trying to drive service technician productivity, you could modulate hiring of new service technicians. You could reallocate development resources to align better with your growth models—keep in mind there is a lag between development investments and returns. There are a number of levers depending on your specific business that you can apply to help your business adjust and change behaviors.

3. **Make organizational changes to deliberately direct the decline/growth curve**
   Finally, any change you are trying to make in your business will face resistance and limited results until you institutionalize them through your operating model and organizational structure. This can get complicated because you have to consider many things before making any drastic changes—how much of the current channel access do you need to promote your IoT business, how does the IoT products and services interface and integrate with your existing products, what new capabilities are required in your business delivery and supporting IT infrastructure to realize the potential of IoT-based products and services, how different incentive structures and motivators play with the current fabric of the organization, and so on. While considering all of these important questions, you also need to assess the leadership profiles and approaches required to make your new initiatives and transition successfully. You have to balance the entrepreneurial traits with broader organizational skills. Once you complete your analysis, you need to plot your potential actions and implications against the growth/decline curve. At this point, you need to examine how your actions and implications influence the gradient and direction of your curves.

14.7 Summary

Any dimension associated with IoT carries a big promise. In the same spirit, according to some recent studies, the IoT monetization market is expected to grow from USD 23.09 Billion in 2015 to USD 443.27 Billion by 2022, at a CAGR of 52.98% between 2016 and 2022. Whatever number you choose to believe, the impact of IoT is big [4].

The purpose of any business is to create economic value for shareholders. So effective monetization is key to the success of any effort. In this chapter we have explored the various nuances which IoT brings to the discipline of monetization.

References

1. (Fenn 2008) Fenn, J., & Raskino, M. (2008). Mastering the hype cycle: how to choose the right innovation at the right time. Harvard Business Press.
2. (Sinha 2014) Sinha, S. (2014). Making Big Data work for your business. Packt Publishing Ltd.
3. (Christensen, 2003) C. M., & Christensen, C. M. (2003). The innovator's dilemma: The revolutionary book that will change the way you do business (p. 320). New York, NY: HarperBusiness Essentials.
4. (McKinsey 2015)McKinsey, http://www.mckinsey.com/business-functions/digital-mckinsey/our-insights/the-internet-of-things-the-value-of-digitizing-the-physical-world

Chapter 15
Building a Winning Team

Like any new business venture, IoT domain also requires you to bring together an awesomely talented high performing team to be successful and win decisively. As we have discussed several times in this book, there are disruptive and transformative elements of IoT. So the criticality of having the right team is as critical as having the right offerings, platform, and business model. In this chapter, we shall discuss the human aspects of building a winning IoT business. While each IoT initiative is different, there are some common elements and best practices that you can benefit from. You will find some similar concepts and descriptions in Chap. 5 of our book "*Making Big Data Work for Your Business*".

In this chapter we shall cover the following topics:

1. Identifying what is different about IoT businesses in the context of people and organizations
2. Understanding the different capabilities you need to bring together to build a successful business
3. Organizing your teams for success
4. Recruiting the best talent for your business
5. Motivating the team to achieve the awesome results you desire
6. Working collaboratively with partners
7. Tapping into the educational ecosystem

Human resource is a much-evolved discipline. There are many successful practices around building successful organizations. Our attempt here is to supplement known wisdom with some elements unique to the IoT domain.

S.R. Sinha, Y. Park, *Building an Effective IoT Ecosystem for Your Business*,
DOI 10.1007/978-3-319-57391-5_15

15.1 Identifying What Is Different About IoT Businesses

As we have discussed in the preceding chapters, many things are different about IoT. Here we will focus on some unique dimensions from the product development and marketing/product management standpoints.

15.1.1 Product Development and Support

In nearly every IoT business, hardware, software, communication, and, cloud technologies are usually integrated into a single product facing the end customer. You could be a player in the value chain of the building, delivering or supporting the product, but usually, the product is a combination of all the factors we just mentioned. In more traditional businesses, these disciplines necessarily do not always converge. In fact, there are separate industries and disciplines dedicated to these individual aspects. For success in IoT, you need to understand the leverage the convergence very well. Multi-disciplinary development and support are difficult because our current education systems train us to think in silos.

Due to its evolutionary and novel nature, most IoT businesses require a very entrepreneurial approach. Even inside of a large corporation you need to become an entrepreneur [1]. Developing and maintaining such an environment requires heightened skills or curiosity, experimentation, dealing with uncertainties, rapid prototyping, iterative and agile development, customer centricity, amongst other similar traits. While this can be a lot of fun, it can also become a very stressful environment and people involved must be able to deal with the pressures. You also need multiple capabilities to be self-contained within the team [2]. Sometimes the numbers of people from different skill backgrounds may not be large. So you need a very collaborative environment. People must also naturally be able to reach out to their peers outside of their team—into the broader corporate set-up, other industries, and companies, academic institutes to learn and leverage.

All IoT businesses are very dynamic. We are still in the formative stages of the domain. So people must be able to adapt and respond very quickly to changing the environment. You need processes and predictability mechanisms, but they cannot override the needs for agility.

More than most other environments, user experience (UX) and operational excellence (OpEx) play a disproportionate role in the success of IoT businesses. Features, functionality and business value are still very important, but UX and OpEx create more differentiation, attracting new customers and use cases. You not only need those skills in the team, but everybody in the team must have some decent level of appreciation for these disciplines.

Traditionally most products and offerings are built for longevity [3]. You need to able to design IoT products and offerings for quick refresh cycles to accommodate for changing user needs, technology, business models, and industry trends. Agile development methodologies like Scaled Agile Frameworks help. While developing you need to think of extreme modularity and adaptability. This requires paradigm differential thinking in design and platforming.

The nature of IoT products and applications has furthered miniaturization of both hardware and software. Miniaturization of electronic components have been underway for a long time, trends in hardware and software are more recent. While sizes might be coming small, expectations around functionality and usability are increasing which requires you to pack more into same or smaller footprint. You also need to design for simplicity to increase appeal and adoption. Skillsets required for such a development environment are different.

15.1.2 Product Management and Marketing

As this is a new domain, you need rapid demand creation and you need to do it all the time. Constant strategizing on reaching new customers, gaining their mindshare, and driving adoption of your products is critical to success. Given the dynamic nature of the space, traditional concepts around long marketing and product management need adaptation [4].

You also need a perpetual engagement with your customers to protect and grow your install base. This requires you to innovate swiftly around new reasons and ways to communicate, create new value streams for customers, and increase their loyalty.

In many IoT applications, the threats of substitution are high. So you need a stronger focus on brand and value proposition. With the proliferation of multiple social media outlets and their influence on customer perceptions, brand and value proposition needs constant vigil. Normal methods of feedback and response are inadequate. Similarly, marketing models have to be adjusted to cater to these new methods of customer engagement.

In the IoT space, often you go to market through multiple non-traditional channels. Entry barriers are relatively low in IoT-based offerings. With multiple options to service customer needs, channels will have less propensity for loyalty. You also need capabilities around quick feedback and closing the loop with channel intermediaries and customers. All this necessitates more active channel management than ever before.

15.2 Understanding the Different Capabilities Required for Offering Development

Now let us understand in greater details some of the different disciplines that are critical to the success of your IoT business from an offering development standpoint. Here we have only covered those that have any specific IoT connotation. Generic skills around project management, quality, etc. have not been addressed here.

15.2.1 UX Researcher

UX researcher engages in understanding customer needs, wants, behaviors and translating them as inputs to the design process. This is different from normal market research methods because you need the ability to understand the psychology and behaviors of individual users and synthesize them into common themes of design. Good research does not rely on proving broad hypotheses but works more on uncovering latent wants and preferences. Good UX goes beyond normal survey methods into more interactive and iterative ones. UX research for IoT will require familiarity of the researcher with underlying technology and opportunities to be able to translate user feedback into meaningful offering design.

A good UX researcher needs strong background on investigative methods to look beyond the obvious especially when working on breakthrough ideas. S/he should be able to empathize with users and do sentiment analysis in order to understand behaviors and what drives them. S/he must be able to balance quantitative and qualitative research methods and must have excellent articulation skills to communicate the interpretations. S/he should be great in collaboration with the other product development teams. They play a key role in being the first interface point with users and influencing the course of how products and businesses are designed.

Many leading universities like the University of California at San Diego or Berkley offer specialized user research courses. There are numerous online courses and resources that can help too.

15.2.2 UX Designer

UX designers take the inputs of the UX researcher and translate them into designs for actual products and offerings including interaction design, information architecture and flow, and specifications for hardware and user interface (for interactive software) design. UX designers will create personas for the products and its usage, identify the key features that users care for, and define how they might want to use those features. UX design has to satisfy the purpose of the product; UX designers play a pivotal role in aligning and achieving the product vision. UX design will develop the storyboards for the product's usage by different persona. During the course of the development, UX designers will work with engineers to test for usability.

UX design is a very established discipline and many schools offer courses on this topic. Many UX designers have a background in fine arts, architecture, or psychology. Engineering and technology orientation always helps with UX design. They always keep the customer in the forefront and use the product to create delightful experiences. Good UX designers are very creative individuals who can be very innovative. While many creative disciplines can tolerate loners, UX designers have to be social. They will be relentless about promoting new ways of thinking and working, championing design led thinking through the rest of the organization. They need be collaborative and be able to influence engineers and other development team members into creating aspirational products.

15.2.3 Industrial Designer

The role of an industrial designer has been described to be one who is to create and execute design solutions for problems of form, function, usability, physical ergonomics, marketing, brand development, and sales. Industrial designers are specialists who focus not only on design but manufacturability and maintenance of the product. They bring many disciplines together—design, engineering, material sciences, manufacturing processes, and marketing. Industrial design has to not only account for the physical product design but also address the needs of software interactivity on IoT devices. Consequently, they need to understand display methods and technologies quite well. This is very important because different display technologies have inherently different physical characteristics which need to be tackled individually, e.g. designing for transparent vs non-transparent requires different approaches to display bonding, surface design and assembly method at the factory. They have to address the miniaturization and adaptability needs of IoT products and allow for future SKU variations with minimal impact to the core design.

Compared to UX architects, industrial designers are grounded more in engineering and manufacturing disciplines. As IoT enabling technologies are fast evolving, they need to study well to keep up the new possibilities. Their collaboration networks extend beyond users and engineers into the entire supply chain that brings the product together. For example, Industrial designers have to work with local tooling vendors, plastics/film suppliers, prototype shops and other suppliers who will help create a number of reference models for review during the product development phase. Good industrial designers are deft with sketching and also using design tools that create artifacts for engineering and manufacturing. Their aesthetic senses have to be exemplary. Hands-on experience with development help industrial designers appreciates the practicality of their ideas.

15.2.4 Technology Researcher

This is a critical capability required for successful IoT businesses to keep up with the changing landscape and most importantly the convergence associated with technology, business and customer experience as applicable to IoT businesses. There are well-established practices around advanced research, they go deep into a subject. IoT technology research requires multi-disciplinary approach into advanced analytics, data management, communication technologies, cyber security, software engineering, and embedded electronics in addition to the domains that the specific IoT product or offering is trying to address. Technology research requires frequent interaction with and deep dive into competitive products, adjacent domains that might influence your business and technology, suppliers. Technology research must guide you in making strategic decisions with a technology context. We have seen several IoT initiatives achieve less than a desirable success due to poor technology choices that lead to increased costs and reduced capabilities.

IoT technology researchers are innately curious people with a strong academic and research orientation. They should preferably come from a fundamental sciences background and have deep expertise in applied engineering. They should be able to understand the implications of forwarding moving technology for your specific IoT-based businesses. For example, a communications expert will be able to comprehend what are the differences between 5G and other networks, but your technology researcher must also be able to contextualize the consequences for your business. They are typically at the pinnacle of their knowledge and expertise. Successful people in this profile are not motivated by the adrenalin rush of quick hit public accomplishments, but draw solace in methodical pursuit of knowledge acquisition and sharing. It is good to have some research capability most adjacent to your problem domain on staff, but be sure to augment this capability with a network of other researchers, academics, and experts.

When working on challenging schedules and targets, we often get the temptation of seconding all talent on to launching products and solving customer problems. In our experience not having dedicated technology researchers with no tie-ins to specific products and programs is critical to long-term success of your business, somebody needs to watch out for future disruptors.

15.2.5 Data Scientist

Data scientists play a key role in your IoT development. They bring together the possibilities and capabilities of statistical methods for analytics, computing and domain knowledge together to solve business problems using data as a backbone. They take a multidisciplinary approach to your ecosystem and product development. As an integrated discipline, data sciences is relatively new and has evolved over the last 10 years. With the advent of big data technologies, this has come to the forefront. They work out the data management, processing and analytical components of your IoT ecosystem. They will often develop proof of concepts independently before broader teams get activated. Dr. Andreas Weigend, Head of the Social Data Lab at Stanford and former Chief Scientist, Amazon.com has very succinctly put into context the significance of data scientists. He has said "We live in a data-driven world. Increasingly, the efficient operation of organizations across sectors relies on the effective use of vast amounts of data. Making sense of big data is a combination of organizations having the tools, skills and more importantly, the mindset to see data as the new 'oil' fueling a company. Unfortunately, the technology has evolved faster than the workforce skills to make sense of it and organizations across sectors must adapt to this new reality or perish."

Data scientists must have strong knowledge of advanced mathematics and statistics. They are very familiar with modeling techniques like regression analysis, time series analysis, Bayesian networks, constraint theories, game theory, Weibull analysis, and other algorithmic approaches. They too must have the ability look beyond the obvious and find relationships through data. They are very good with data modeling, processing, storing, and mining. They need have an extensive understanding of artificial intelligence, machine learning, and natural language processing; most IoT products and ecosystems have elements of all of them. Mostly good data scientists have in-depth knowledge of programming with Python, R, C, C#, Java and other popular languages.

Data scientists are artists with data—they are inquisitive about data, can spot trends, think data represents ultimate beauty and can tell stories with data. They think in term of algorithms and can paint a compelling picture of how you need to take a fresh perspective of your business using data and analytics. However, they are not esoteric research scientists; they are very sociable human beings who can translate the complexity of your business and its underlying data into powerful yet simple insights and recommendations.

While still in short supply, it is becoming easier to find data scientists. The educational ecosystem is developing. You can find data scientists from mathematics and computer sciences background. Good data scientists have post graduate and often doctoral studies background with strong academic orientation. We have seen good data scientists coming from other disciplines beyond computer sciences like biology, chemistry, neurosciences, astronomy and other similar science-oriented disciplines; they all use a lot of data. This is a critical role you must have on staff and can augment with people from academia.

15.2.6 Platform Developers

Platform developers bring the IoT ecosystem into reality. They build and maintain the communication interface, device management, data hosting, customer account management, device/user authentication-authorization mechanism analytical processing platform, and the backbone required to drive your IoT business. This too requires a multi-disciplinary and multi-skill approach because of technology convergences. Platform developers have to design and develop for high scalability and rapid adaptability of the product offerings built on their platforms. Platform design also needs to be easily transportable so that you can take advantage of new hosting, device management, security, licensing and other technology opportunities; so the platform developers must be able to componentize well. Their efforts have a significant influence on customers' operational experience from the products.

Platform developers must have expertise in cloud technologies, distributed computing, data management, and communication technologies. The cloud technologies context to these dimensions is key because of the handling of data and compute is different in the cloud with elements of extensive virtualization, content delivery systems, and network management. Many of your platform developers will have evolved from traditional client-server based infrastructure groups with experience in cloud computing. Background and experience in modern programming methods, interfacing with multiple systems, developing APIs and services and secure development methods are useful to be a good platform developer. Platform developers should also have an appreciation for data sciences so that they can help design the infrastructure for most optimal algorithmic execution. They need to be able to collaborate effectively with IoT application and other business support system developers. The platform is critical to the long-term sustainability of your business, so it is best to have this group in-house as much as practical.

15.2.7 Application Developers

IoT applications have a very wide range—from simple home appliances to large complex industrial systems. Application developers are ones who build interactive software offerings for your IoT businesses. These might be embedded into the physical product or instantiated independently through a browser in any type of stationary or mobile device. In more traditional context, embedded software developers and other types of application software developers have belonged to different disciplines; in the case of IoT, they largely converge. Same goes for user interface and data management. IoT application developers have to be excellent generalist programmers and testers; they must be able to work relatively independently. They must also be capable of learning about and developing applications on other people's platforms to reduce effort and get to market quicker. IoT application development usually involves widespread edge computing which requires specialized skills in handling data transport, high volume data egestion, localized control and decision making, security and miniaturized onboard computing. A lot of IoT application development uses standard frameworks instead of writing every single line of code. Usually, IoT applications will have connectivity first and mobile first as core underlying philosophies. In addition to the older programming languages like C, C++, Java, Java Scripts, newer languages like Python, Go, Rust, Parasail, B#, and Forth are also becoming popular for IoT application development; they each have specific implementation advantages.

Application development skills, in general, are similar to any technology environment. Good IoT application developers will know how to work with different operating systems. Like platform developers, they also must be able to design for modularity and transportability. They must have a good grounding on the domain your business is into to be able to reflect the realities and possibilities through the product offerings. They also must have a wide knowledge of integrating with other systems and devices. It is critical that IoT application developers are very conversant with agile methodologies and DevOps processes. Given the compressed schedules and rapidly changing environments, they must be very quick and creative. It should not be very difficult to find good IoT application developers if you know what you are looking for.

15.2.8 Hardware Engineers

IoT hardware is usually different from other products on two counts—they have embedded communication, compute, storage capabilities on the edge device, and miniaturization. These necessitate some different type of hardware engineering skills beyond the knowledge of traditional mechanical and electrical engineering because now you have to account for the computing and connectivity needs of the product. IoT hardware engineers need to understand multi-layer PCB design, different communication protocols, and wireless technologies. Since most IoT devices have an array of sensors, they need to understand their characteristics for optimal footprint, placement, and compensating for interference factors like radio noise and heat rise impact. Reliability takes a new dimension because of low serviceability of most IoT devices. Power management and battery configuration are crucial topics in IoT devices because in most cases they may not be connected to a constant supply electric power source. Traditionally most hardware engineers restricted themselves to the physical products; since some level of software is present in most IoT devices, some level of software understanding also important for hardware engineers. The need to offer a remote application or OS patches and upgrades requires the hardware engineer to work in close collaboration with the software and platform developers for creating a seamless user experience. Like industrial designers, hardware engineers also need to have a very deep appreciation for manufacturability. Most IoT devices will have displays, so integrating different display technologies, bonding displays to the physical product and establishing communications between the display and the PCB are critical skills to know about. Choices made by the hardware engineer have huge implications on product cost in the short term and the ability to scale the device portfolio in the long run; such impact makes this position very critical to an IoT product development program.

Best places to look for good IoT hardware engineers are manufacturers of a mobile phone, controls system, high-end automotive electronics, and medical devices. Electronics manufacturing services (EMS) and original design manufacturers (ODM) are also good places to get some talent from. You do not necessarily need all hardware development capabilities in-house, but having some key staff with broad capabilities will help you navigate the supply chain.

15.2.9 Compliance Specialists

There are many dimensions to compliance in the world of IoT. You have the traditional compliances around quality which have different standards for various device application industries and sometimes geographic jurisdictions. With IoT, you have to often deal with compliance around electromagnetic interference for which standards are established by different countries like FCC for the US, CCC for China, VCCI for Japan, CE for European Union countries, BIS for India, and other similar ones. You also need to build your products to comply with cybersecurity standards like those published by National Institute of Standard and Technology for ensuring data security and privacy protection. Safety and reliability is a big issue with many IoT devices especially those dealing with health and well-being; standards for such scenarios are different and quite stringent. Many of these standards are evolving quickly to keep up with risk exposure factors. So you need dedicated focus on compliance to be successful in your IoT business; more so because you need to meet the obligations of these multiple converged compliances.

Compliance specialist often have a technical background and sometimes a legal one; in either case, they must have an appreciation of the other. They must have a detailed understanding of the various standards you need to comply with and familiarize themselves with how your products comply with them. They should have networks and methods to keep up with the changing regulations. Good compliance specialists must be thorough and diligent to mitigate your business risks. It is good to hire people in this role with experience in agencies either designing or implementing these standards.

For successful and on time product launch, it is critical that you schedule pre-compliance activities based on your past experience with the agencies. Use of standard contraptions that have been used in other 'compliant' products can help with a quick turnaround in certification. A number of agencies have partnerships with third party laboratories that offer such services; we suggest you engage with them early as you work through the design process.

15.3 Organizing Teams for Success

While building IoT businesses, you should think beyond traditional organizational structures with fixed roles, responsibilities, authorities; you need to design an operating model which defines the battle rhythm for an emerging fast moving business. On the surface many of the functions will look similar, the differences lie primarily in accountability and mutual interactions. You need more collaborative approaches and autonomy in decision making for IoT organizations to be successful.

15.3.1 Functions

Following are the key functions that you need represented in the team:

1. **Strategy**—This function is responsible for developing a winning formula for the business which describes economics of the industry, value capture points across customer engagement lifecycle, opportunity landscape—market segmentation and sizing, key customer needs, market and technology trends, competitive landscape, your right to win, profit pool analysis, impact to your other businesses

(if applicable), offering portfolio with customer value proposition and pricing models, and 5-year pro-forma P&L. While this function is not responsible for the technology side of your business but needs to understand points of differentiation created by IoT technologies and be able to track their evolution curves. This function will also maintain your IP portfolio to define your strategic advantage along with your legal team. As required, this function will lead business development through partnerships and inorganic routes.

2. **Marketing and product management**—In traditional businesses, these two functions may exist separately, but in IoT businesses, they are quite intertwined. The products define the market and your go-to-market strategies have a lot of influence on the product. This function is responsible for leading the efforts to capture and increase customer mind share and wallet share. IoT products constantly transform by enabling new capabilities, so lifecycle management becomes quite different than relatively fixed function capabilities of traditional product or service businesses. In IoT, marketing is often delivered through the product itself and does not rely entirely on external channels like traditional businesses.
3. **Sales and channel engagement**—This function is responsible for managing the revenue growth for your IoT business. Since you will frequently go to the market through multiple channels, this team will manage the various channel relationships, understand what is critical to their success, and how best your IoT solutions can enable channel success. Traditionally sales are more about demand fulfillment, in IoT, it has more responsibilities for increasing customer engagement—directly and through channels.
4. **Platform development**—This function builds and maintains your core IoT ecosystem as described in the earlier chapters. This team will have many skills listed in the previous section's '*platform developer*' portion. If you choose to use somebody else' platform, you need a skeleton group to manage that relationship and ensure the ecosystem is able to support your changing needs.
5. **Offering development**—This is the classic product or offering development organization which should include all the capabilities except platform developers from the previous section. Sometimes UX researchers and designers are more closely aligned with any may even belong in the product marketing function. We have also seen instances where technology research and data sciences sit in the platform development organization, another very valid choice. You have to make these choices based on what serves your business best.
6. **Business operations and customer support**—In IoT, your entire business resides on a technology platform. The business operations and customer support function for IoT businesses is responsible for maintaining the customer engagement with your technology and products on an on-going basis. This is where customer's value capture loop is closed. The success of this organization defines your sustained revenue and profits.
7. **Functional support**—These are the traditional functions of HR, finance, legal, procurement, manufacturing, etc. which are not very different for IoT-based businesses; they just need to have a better appreciation of the product, technology, and market differences. For very old traditional organizations, their processes might have to be tweaked to meet the agility needs.

Make your team a self-contained autonomous unit if you are part of a large organization. This business is different than others, so give them the freedom to perform and have the accountability for results. If you are a start-up, this aspect is not an issue, but do manage to negotiate some leeway with your investors.

15.3.2 Review Cadence

In traditional businesses, we usually have monthly and quarterly reviews around market growth, trends, business performance, customer acquisition, competitive actions and gap closure initiatives if the business is not meeting its goals. For IoT businesses, you need a bit more. You need to monitor the growth of your connected install product and customer base, utilization of your products' or any other consumption type metric, and customer usage feedback. You need to pay more attention to the leading indicators in case of IoT so that you can quickly spot trends and respond to them. You should also be hawkishly looking at current competition and what could become one tomorrow? Your reviews should be more frequent and in-depth focused on the 'why' just not the 'what'. The primary orientation of reviews for IoT businesses has to be learning and adapting.

From an internal perspective, you also need to review talent development, key talent gaps and collaboration efforts on a frequent basis.

15.3.3 Leadership Attributes

The leader of the business plays a very vital role in building out your business. It is important that they demonstrate the following attributes:

1. **Passion**—S/he will have to motivate a group of people into venturing into a new space and achieve awesome results. With infectious passion, the leader can drive the team for high performance.
2. **Team building**—Multiple different skills and capabilities need to converge to deliver the promise of IoT. We have discussed this before. The leader must be able to integrate divergent highly capable people into a cohesive unit. Being able to build great teams is critical for success.
3. **Navigation and alignment**—If you are using IoT to transform existing businesses, the IoT team will need to collaborate with other businesses for shared successes. This requires higher order navigational skills within large organizations which can be very rare to find. The leader must also be able to align multiple groups with different priorities into one common theme and action plan. Being collaborative helps this cause.
4. **Convergence**—Any IoT business has multiple convergences and transformation points. The leader must have an understanding and experience managing groups which are impacted by the convergence.

5. **Technology oriented service orientation**—IoT success depends on the ability to combine technology, solutions, and services wrapped around those. The DNA for this is very different from traditional product oriented business. Working knowledge of how engineering, IT or knowledge processes services industries work will be very helpful.
6. **Channel management**—IoT opportunities will be realized through a combination of multiple channels—company's direct access to the market, other distribution methods, technology companies, independent software vendors (ISVs) and other possibilities. Having some personal experience in building and/or running such channels helps accelerate market reach and adoption.
7. **Credibility**—A successful leader needs to have some level of personal credibility. Serving on expert panels, advisory boards, speaking at conferences, being published and interviewed, all of them count. You will be surprised at the extent of research customers and your future employees do to understand what you can bring to the table as a leader.
8. **Entrepreneur**—The leader must demonstrate skills and attitude of an entrepreneur but know how to work inside of larger organizations.
9. **Visionary**—It always helps if the leader of the team, in addition to being good at everything else, is also the one championing the vision for the IoT business. If s/he is not a visionary but a great leader otherwise, they must have the ability to be the principal cheerleader for the vision.

15.4 Recruiting the Best Talent

Recruiting for any new emerging space is always challenging. Doing so for IoT is particularly complicated because of the unique skills available in short supply. Rapidly evolving technology and business model convergences make this even tougher. You need the best talent who know they are desired. Here are some ideas we have seen work in this endeavor:

1. Create an aspirational vision for your business. Use this to not only motivate but challenge the potential recruits. While ambitious, your vision must be believable and practical, otherwise, there will be limited credibility behind it. You must be very articulate and descriptive about the vision. You must also be able to communicate why this is achievable and your company must lead the way. Inspire people to effect change in people's lives in a way that seems like an obligation, because you and your team which includes the potential hires are best suited to deliver the promise. This will help talented people find meaning in their work and help connect with your goals.
2. Employ your best and brightest in attracting talent. When we look at hiring some promising data scientists, Dr. Park, an esteemed data scientist himself personally gets involved. The candidates get the assurance that this is a place where they will get the opportunity to learn from the best and sharpen their skills. If you are not hiring in bulk, this is actually a quicker and more effective method than relying on traditional filtering and recruiting methods.

3. Leverage others to influence your hiring pool. When recruiting from universities, we often engage the faculty very directly and deliberately. They have the trust and respect of the students (make sure first that is the case through some secondary research), so the students are more likely to join your initiative if recommended by the faculty. When recruiting laterally, explore such influences with folks from other technology companies, analysts and industry companies whom your targeted candidates may have good interactions with.
4. Generate leads through your current teams. We have seen this work very well all the time. Your best people know great talent elsewhere, perhaps they have worked together in the past or networked for some purpose. These relationships can be exploited to expand your team in a reliable manner; your team will not make the recommendation unless they are sure of the candidate's credentials after all their credibility is also at stake.
5. Use firms specializing in IoT recruiting. In the world of executive hiring, we have seen recruiters often having senior executive background themselves, this helps understand candidates better and establish credibility. Similarly, we are slowing seeing the emergence of agencies specializing in recruiting for IoT. This allows them to have the appropriate technology and business model background. Focus also allows such recruiters to network better.
6. Share some form of incentive in the form of stock options or restricted stock units linked to the business performance of your IoT initiative. This drives more ownership mindset amongst the candidates and opens avenues for them to be financially rewarded very well if things work out. For start-up companies, which are many in the space of IoT, this is a frequently employed route; it helps get great talent in an affordable manner and promote loyalty. For larger corporations with well-established rules this type of approach might be difficult but work with your HR partner to explore possibilities.
7. Create executive connect opportunities for people with rare skills. Such people know their value and will appreciate the courtship. This also helps your executives better appreciate the lengths to which you are going in making the business successful as well as develop an appreciation for the uniqueness of the space.
8. Clarify your stringent standards for people. This sets the bar high and makes talented competitive people want the job more. However, you have to be very careful when employing such tactics not to be seen as arrogant and elitist; that could turn people off.
9. Assure people on your flexibility and intent to give them new experiences. People working in the IoT space do not want to get bogged down in traditional silo environments. If they know that completing their assignments successfully enable them to learn and do new things, they will be more inclined to join you. Show your potential hires career maps which demonstrate real case studies of how people have grown in capabilities and responsibilities over time. If you do not have history to show for, paint a picture for future possibilities.

Recruiting the best talent is never very easy. You need strong persuasion and perseverance for success.

15.5 Motivating the Team

If you recall from the first section of this chapter, there are few differences between normal businesses and IoT-based ones. Chief among them are the fast paced, dynamic and technology rich environment that are the cornerstone of any IoT business. Your strategies to keep the team motivated and high performing have to be adapted to suit such an environment. In our book '*Making Big Data Work for Your Business*' we had outlined a few attributes which differentiate the right people working on big data projects, we have seen the same to hold true for IoT initiatives:

1. They are not bothered by organizational hierarchies, even though they will show respect for authority and structure
2. They do not get excited most by normal motivational incentives like financial or positional rewards
3. Might be reticent at times, but they like to share and collaborate, and be generally helpful to others in the right environment

Here we share some of the best practices we have practiced ourselves and observed elsewhere:

1. Incentivize people on the growth of your connected install base in addition to normal business impact and customer experience parameters as your business moves forward on a growth trajectory. Most incentive schemes are based on simplistic revenue, profitability, productivity and customer satisfaction goals. While these are important for every business, in IoT ones, there will be a lag before you see the significant impact of those parameters. On the other hand, if you can quickly grow your installed base of connected products and customers, you have increased your chances of long-term success. Incentivizing the growth of your connected install base will encourage your teams to ideate on novel methods of creating customer value, increasing adoption and improving stickiness; given the dynamic nature of IoT, these are critical for sustained future growth.
2. Provide opportunities for learning new technologies and gaining new experiences. There is a lot of new development happening in this space both in terms of technology and business models. The more people learn, they will be able to apply these better for your benefit. This will also help them be current and keep your business fresh. Send your people to conferences and trade shows, let them learn from competitors, suppliers, and adjacent industries. Even if you operate in industrial internet space, participate in consumer electronics shows; there is the more voluminous adoption of new technologies there. IoT conferences are another good place to see new use technologies and applications. Organize training or send your teams outside for training to upskill them. Your people will appreciate that you are unleashing their potential.
3. Allow for experimentation. Traditional business management principles are designed to ensure higher degrees of predictability in every aspect—be it product

development or sales or customer service. These are important, but a lot of exploration is required for differentiating and disrupting using IoT. So you need to allow for some level of experimentation with your team. You must build such allowances for innovation in your business model and plan design. Most successful IoT businesses have been driven by innovation and adapted based on learnings.

4. Create opportunities for demonstrating thought leadership and technology excellence. Your team believes that they are working on cutting edge technologies and businesses that are changing the world. Feed their feeling by encouraging them to share their thought leadership and technology excellence with the rest of the world without of course compromising any competitive advantages. There are several ways of doing this—filing for patents, publishing white papers or case studies, and speaking at conferences. These are relatively low-cost methods. They also help with your company or institution's branding and marketing efforts. Patents also contribute to your organizational valuation. Provide financial incentives by sponsoring such participation in patents, PR, and publications.
5. Enable learning firsthand about the impact of their efforts on customers and channel partners. Let your team walk in the shoes of your customers and the various routes your products reach the customers. They will develop a much better appreciation of the problems they are solving or can solve. This will trigger many new ideas and possibilities. Do not leave this exercise only to UX researchers and designers. In traditional businesses, such activities are limited to select few from product marketing. When markets and businesses are well established, you can afford to have limited experiential learning; for evolving markets and opportunities, experiential learning is existential.
6. Reward collaboration. Multiple skills and efforts need to converge in order for a successful IoT business to emerge, we have discussed this extensively so far. So you need to encourage and reward collaboration within the team and across boundaries in your organization. Sometimes your team will reach out to the broader ecosystem and find unique capabilities that can add to your offering portfolio; appreciate those efforts. If people express a desire for cross-platform or discipline learning, help them broaden their horizons, they will get better at their core job.
7. Celebrate successes where the impact is felt by other parts of the business. In certain scenarios, the immediate and direct beneficiary of your IoT efforts might be a more the traditional line of business. So you may not see the impact for your group as directly. Even in such cases, celebrate the wins of others which you have helped engineer.
8. Praise if the team makes course corrections and internalizes learnings based on failed experiments. Keeping in line with the spirit of experimentation and innovation, you will likely experience failure on certain occasions. If your team can understand what lead to the failures and modify their approach to avoid them in future, celebrate them. By doing so you will promote a learning organization that behaves responsibly.

9. Do not stifle creativity and speed through organizational hierarchies and archaic processes. This is a sure killer for any entrepreneurial initiative which you need in hordes to build and grow your IoT business. So you need to be careful in choosing your managerial and oversight staff. Often if they lack self-confidence and contextual understanding, they buy time and increase their perceived value by introducing multiple checkpoints which may not be necessarily adding value to the business. Modular self-contained and self-organizing teams demonstrate greater levels of accountability and perform better.
10. Get only the best talent. This has multiple benefits. You get to develop your IoT ecosystem and business faster, potentially realizing business value quicker. Your team will respect the talent and accelerate their learning. They will have a few competitive behaviors and be motivated to always excel, keeping their edge. You want to make an impact fast and possibly first, so you can little afford not to have the best people working on your problems.

15.6 Tapping into the Educational Ecosystem

Often the academic world is early in its knowledge of technology and business practices because of its fundamental research orientation. The same is true for core enabling technologies and practices for IoT. This extends even to setting industry standards. So working with academic institutes can give you early insights on one hand and fill up your skill gaps on the other hand. Such relationships extend your influence network on broader technology direction, customer choices, talent pipeline and standards selections.

There are many ways to engage with academic institutes:

1. **Funded dedicated research programs**—If you are not constrained by the paucity of funds and have specific problems to be solved, engaging professors and students in solving these research problems is a very effective approach. This reduces the pressure on your internal team to work the problem. You may even be benefitted by some independent and divergent thinking. You can do these engagements either as a one-time activity or a sustained program with new problems being worked upon as time goes by. Usually, in these engagements, the IP is owned by you because you are essentially paying for its development. If the institute does not have a formal way of engaging in dedicated research with industries, they will generally allow their faculty to design custom engagements or consulting type assignments.
2. **Collaborative research programs**—When funding is an issue or you need other companies to join hands to solve problems, a more collaborative approach is better. It has the same benefits of the former model. One difference will be in ownership of the IP, you may have to be flexible as you are not paying for all of it.
3. **Consortium approaches**—Some institutes have started forming IoT consortiums comprising of multiple companies representing different industries and

technology companies. There is usually a fee to join such forums and you get to benefit from collective knowledge sharing and incremental research activities. This is basically extending your learning and experimentation network. This model is usually low investment but not very good for solving complex problems.
4. **Co-development of courseware**—You can work with an institute to develop courses dedicated for IoT like curriculum for data sciences programs. These can be used dually for internal training as well as developing a better talent pipeline for future. This is a good way to get some branding and buzz around your company too.
5. **Student engagements**—These include the traditional models of internships, summer projects, and recruitment. These are good to get raw talent and getting to know your potential future workforce.

15.7 Summary

IoT creates exciting opportunities for people to work on. There are a number of disciplines that need to figure out innovative ways of working together to build a winning IoT business. In this chapter we have discussed so far:

1. The different capabilities required to build an IoT business
2. How to organize teams effectively
3. Best practices on recruiting, retaining and expanding your talent pool

So far we have covered nearly every aspect of building an IoT ecosystem for your business. In the next and final chapter, we shall discuss some new models to measure the effectiveness of your IoT ecosystem beyond the normal business metrics.

References

1. Benefield, G. (2008, January). Rolling out agile in a large enterprise. In Hawaii international conference on system sciences, proceedings of the 41st annual (pp. 461–461). IEEE.
2. Blank, S. (2013). Why the lean start-up changes everything. Harvard business review, 91(5), 63-72.
3. Kettunen, P. (2009). Adopting key lessons from agile manufacturing to agile software product development—A comparative study. Technovation, 29(6), 408-422.
4. Stark, J. (2015). Product lifecycle management. In Product Lifecycle Management (pp. 1-29). Springer International Publishing.

Printed in the United States
By Bookmasters